AF247436

Multiscale Modeling of the Skeletal System

Integrative approaches to biomedical research promise to advance our understanding of the human body and physiopathology of diseases. In this book, the author focuses on the skeletal system, demonstrating how multiscale modeling can determine the relationship between bone mechanics and disease. Introductory chapters explain the concept of integrative research, what a model is, predictive modeling, and the computational methods used throughout the book. Starting with whole-body anatomy, physiology, and modeling, subsequent chapters scale down from bone and tissue levels to the cellular level, where the modeling of mechanobiological processes is addressed. Finally, the principles are applied to address truly complex, multiscale interactions. Special attention is given to real-world clinical applications: one in pediatric skeletal oncology and one on the prediction of fracture risks in osteoporotic patients. This book has wide interdisciplinary appeal and is a valuable resource for researchers in mechanical and biomedical engineering, quantitative physiology, and computational biology.

Marco Viceconti is a world expert in computational modeling of the musculoskeletal systems. One of the main driving forces behind the Virtual Physiological Human (VPH) Initiative, which aims to develop the methods and technologies to realize the dream of systems biology, systems medicine, and the physiome, Viceconti is currently the Director of the VPH Institute, the international non-profit organization that drives the VPH initiative.

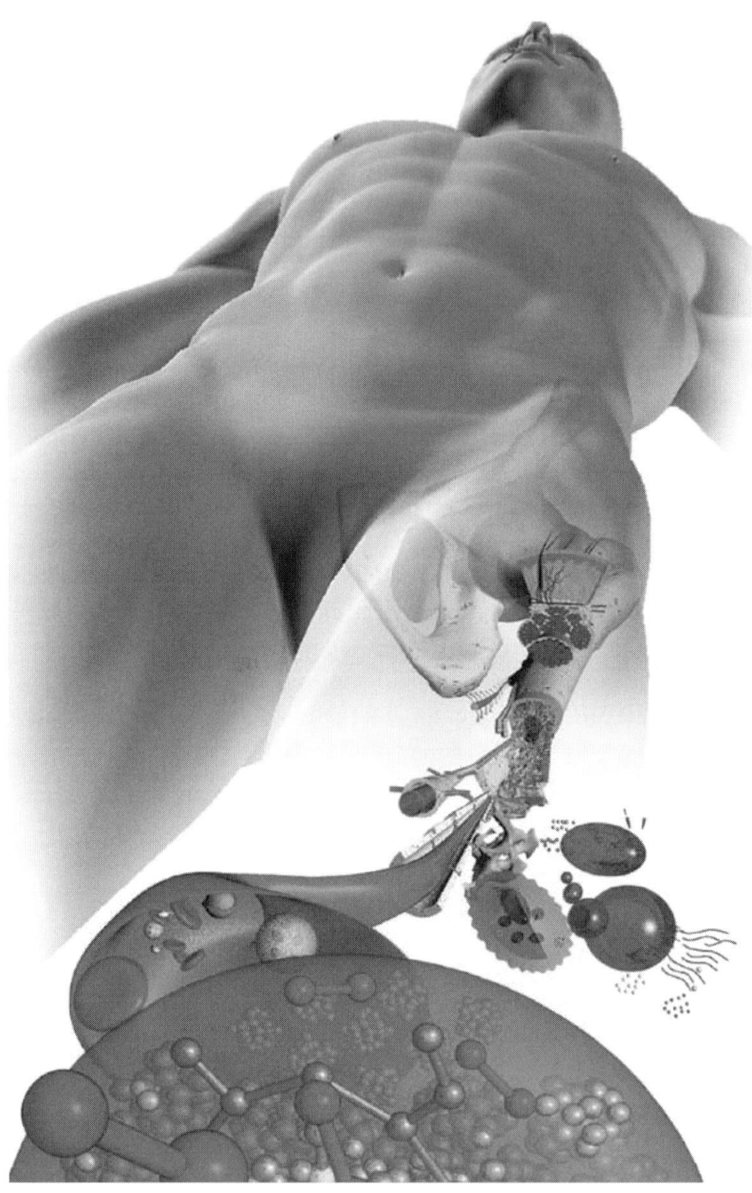

From the Atom to the Whole Body. Luigi Lena and Marco Viceconti. © 2011 VPH Institute, reprinted with permission.

Multiscale Modeling of the Skeletal System

MARCO VICECONTI

Istituto Ortopedico Rizzoli, Bologna, Italy

CAMBRIDGE UNIVERSITY PRESS
Cambridge, New York, Melbourne, Madrid, Cape Town,
Singapore, São Paulo, Delhi, Tokyo, Mexico City

Cambridge University Press
The Edinburgh Building, Cambridge CB2 8RU, UK

Published in the United States of America by Cambridge University Press, New York

www.cambridge.org
Information on this title: www.cambridge.org/9780521769501

© M. Viceconti 2012

This publication is in copyright. Subject to statutory exception
and to the provisions of relevant collective licensing agreements,
no reproduction of any part may take place without the written
permission of Cambridge University Press.

First published 2012

Printed in the United Kingdom at the University Press, Cambridge

A catalogue record for this publication is available from the British Library

Library of Congress Cataloguing in Publication data
Viceconti, Marco.
   Multiscale modeling of the skeletal system / Marco Viceconti.
      p. ; cm.
   Includes bibliographical references and index.
   ISBN 978-0-521-76950-1 (hardback)
   I. Title.
   [DNLM: 1. Models, Biological. 2. Musculoskeletal Physiological Phenomena.
   3. Biomechanics – physiology. 4. Computer Simulation. WE 102]
   612.7–dc23   2011038239

ISBN 978-0-521-76950-1 hardback

Cambridge University Press has no responsibility for the persistence or
accuracy of URLs for external or third-party internet websites referred to in
this publication, and does not guarantee that any content on such websites is,
or will remain, accurate or appropriate.

A model is an invention, not a discovery
Massoud, T. F., *et al.* (1998). *FASEB J*, **12**(3), 275–85.

To Paola and Anna, dancing at the *Paradise*.

"Professor Viceconti is one of the pioneers of the integrative approach to biological systems and his book is a beautiful illustration of this approach. Beautiful in all senses: sound intellectual argument, clear informative illustrations, fascinating practical applications, and it forms a sustained development of a field of research that is fundamental to the success of any attempt to create the Virtual Physiological Human."

Denis Noble, University of Oxford

"Marco Viceconti is the most energetic advocate of the VPH concept – the use of standards-based multi-scale modeling techniques strongly coupled to data at all levels and, wherever possible, healthcare applications. His speciality is musculo-skeletal modeling and in this book he brings a unique combination of technical expertise (including modeling strategies and numerical methods), clinical experience and modelling philosophy, always guided by the pragmatic problem solving approach of an engineer. The book is a delightful combination of philosophical musings, personal anecdotes, scientific insights and technical advice. It will be of great value to the musculo-skeletal research community, particularly those interested in the clinical application of multi-scale modeling."

Peter Hunter, University of Auckland

Contents

Preface

The human skeleton has been intensively investigated from the very beginnings of biomedical science, and still remains a central research topic for many scientists worldwide. As the skeleton primarily has a mechanical function, it is reasonable that a vast number of these researchers look at the biomechanics and the mechanobiology of the skeleton. These investigations are pursued at many different scales, from the molecular level to the whole body, but in the last few years there has been a growing awareness that many of the important mechanisms that we observe are the result of complex systemic interactions that occur at different scales. This forces us to invent new ways of modeling the mechanobiology of the skeleton, ways that let us overcome the limits of a reductionist approach and account for the complexity of these systemic interactions.

The aim of this book is to provide the reader with a systematic update on where we are in terms of modeling the biomechanics and mechanobiology at each characteristic scale, and to show in which directions I believe it is possible to pursue that integration across the different scales that essential research now demands.

In his *L'ordre du discours* (Paris: Gallimard, 1971), Michel Foucault writes: "Within its limits every discipline recognizes true and false propositions; but it pushes outside its limits a whole teratology of knowledge." Foucault believes that the monsters of science lie at the outer limits of the bodies of knowledge of each discipline: but it is in these border regions, where different bodies of knowledge nearly touch each other, that the most exciting discoveries are always made. I believe that each scale delimits a corpus of methods, idealizations, even epistemologies. So modeling across scales is very much like exploring the boundaries between different scientific domains: it is a difficult and troubled journey, but the prizes that reward those brave enough to undertake it are huge.

I hope that you will find this book frequently interesting, sometimes entertaining, and only rarely boring. I also hope that reading it will ignite in some of you the curiosity to look at the mechanobiology of the skeleton in a different light, that of integrative biomedical research.

Acknowledgements

The list of people who contributed directly or indirectly to this book is too long to be cited exhaustively here. My three primary sources are the team at the Laboratorio di Tecnologia Medica of the Istituto Ortopedico Rizzoli, the VPHOP Consortium, and the VPH Initiative.

This book celebrates my twenty years of work at the Laboratorio di Tecnologia Medica of the Istituto Ortopedico Rizzoli. During these years I had the opportunity to work with some of the brightest and most gifted youngsters active in our research domain; to all of them a big thank you, all I taught you came back in what you taught me. In particular, I would like to name my closest colleagues and friends: Luca Cristofolini, Fabio Baruffaldi, Massimiliano Baleani, Fulvia Taddei, Saverio Affatato, and Susanna Stea. Without them, much of what we achieved in those years would have been much harder if not impossible. Three post-docs in the computational biomechanics group helped in the preparation of this book, revising and advising: Martino Pani, Saulo Martelli, and Enrico Schileo. Luigi Lena prepared most of the original illustrations in this book.

The VPHOP Consortium is a European team of 19 institutions developing next-generation technology for the prevention of osteoporotic fractures, using multiscale modeling approaches. As coordinator of this incredibly strong group of scientists I had the opportunity to learn a great deal, and much of what I learnt from these colleagues found its way into this book. Of over 200 people involved with the project, I would like to mention in particular Ralph Müller, Keita Ito, Bert van Rietbergen, Bill Taylor, Stephen Ferguson, René Rizzoli, and Karl Stroetmann.

Most of the general ideas on integrative modeling and the multiscale representation of human physiopathology emerged from the "Virtual Physiological Human" initiative, which has been radically transforming the scenario of European biomedical research since 2005. The VPH Initiative today involves over 2000 researchers worldwide, and with many of them I debated some of the ideas exposed in this book, one way or another. Among the others, I would like to thank Peter Hunter, Rod Hose, Jim Bassingthwaighte, Denis Noble, Alex Frangi, and Peter Kohl.

1 Introduction

An introduction of multiscale modeling of the musculoskeletal system, from a general understanding of what a model is, and of the role that integrationism can play in biomedical research.

1.1 Integrative research: a new approach to biomedical research

The *primum movens* of this book is the emergence, some years ago, of a radically different concept of how biomedical research should proceed. Since the seventeenth century, *methodological reductionism* has been the principal and most successful approach to the investigation of nature. By this term we mean, "An approach to understanding the nature of complex things by reducing them to the interactions of their parts, or to simpler or more fundamental things." (Wikipedia, 2008). Although the term and its meaning have been vastly debated, the most common interpretation, which I shall use in this book, is that in order to investigate a complex natural phenomenon we identify its composing parts and investigate each of them separately, pretending that they are independent of each other. Of course, this is not true in general, but in most cases the error we commit in making this idealization is acceptable.

The problem with this approach is that methodological reductionism too frequently becomes *causal reductionism*, "which implies that the causes acting on the whole are simply the sum of the effects of the individual causalities of the parts" (Polkinghorne, 2007). If we can understand how the parts work, our understanding of how the whole works will follow. In genetics, scientists started to investigate the genes, a fundamental component of the cell, in the hope of understanding how information was passed from one generation to the next (methodological reductionism). After a while, various opinions emerged, arguing that the information encoded in the genes is the only and true "recipe of life," a position that led to the so-called *gene-centered view of evolution*, made popular by Richard Dawkins's masterpiece *The Selfish Gene* (Dawkins, 1976).

This excessive attention to the so-called *upward causation* (genes cause the cell's behavior, which causes the tissue's behavior, which causes the organ's behavior, which defines the behavior of our whole body) has for a long time obscured the strong evidence that, in most living organisms, upward causation coexists with mechanisms of *downward causation*. This term, introduced by

Donald T. Campbell, in the context of systems theory (Campbell, 1974), can be summarized as follows:

The whole is to some degree constrained by the parts (upward causation), but at the same time the parts are to some degree constrained by the whole (downward causation).

The interested reader can find an excellent review of all these arguments in Denis Noble's *The Music of Life* (Noble, 2006).

A number of fundamental research questions in modern biomedical science involve physiopathological processes, where such downward causation appears to be very important. If this is the case, we shall never be able to answer these questions fully using the classic reductionist approach. We need a new approach to biomedical research, something so radical that it will change the way we have done research in the last three hundred years. Recently, this new approach has been labeled *integrative research*, to indicate an empirical science approach where methodological reductionism is complemented by an investigation of the systemic aspects involved, which is called *integrationism* (STEP Consortium, 2007).

The idea of integrative research is very attractive, but the question of how this can be achieved in practice remains open. A "physiome project" was proposed in a report from the Commission on Bioengineering in Physiology to the International Union of Physiological Sciences (IUPS) Council at the 32nd World Congress in Glasgow in 1993. The term "physiome" comes from *"physio"* (life) and *"ome"* (as a whole), and is intended to provide a "quantitative description of physiological dynamics and functional behavior of the intact organism" (Bassingthwaighte, 2000). Since its early inception, the construction of such a quantitative description, a typical integrative research exercise, was conceived through the use of predictive models and a systematic multiscale approach. This perspective was strengthened when in 2006 the EuroPhysiome group started the STEP Coordination action, which produced the *European Research Roadmap to Integrative Research* (STEP, 2007). This document describes the principal instrument for the practical realization of the Integrative Research vision: the so-called *Virtual Physiological Human*, defined as "a methodological and technological framework that once established will enable the investigation of the human body as a single complex system" [Clapworthy *et al.*, 2007; 2008; Fenner *et al.*, 2008; Hunter and Viceconti, 2009; Hunter *et al.*, 2010; Kohl *et al.*, 2008; Kohl and Viceconti, 2010; STEP Consortium, 2007; Thiel *et al.*, 2009; Viceconti and Clapworthy, 2006; Viceconti *et al.*, 2008a; Viceconti and Kohl, 2010; Viceconti *et al.*, 2010].

Thus, integrative research will be possible when we are able to build models that accurately predict the human physiological and pathological processes at the various spatial and dimensional scales, and we are able to combine them so as to explore the systemic behavior that is at the basis of such processes. Therefore, the basis of integrative research is the ability to build accurate multiscale models of human pathophysiology. But before we proceed, it is necessary to answer to a fundamental question: what is a predictive model?

1.2 What is a predictive model?

1.2.1 What is a model?

Modeling is one of the most used approaches in the human exploration of nature. Virtually all disciplines make use of models, even domains where the empirical component is strong, as in biology or in medicine. However, we still have problems in answering the fundamental question: what is a model?

The word "model" is one of the most elusive terms we use in science. The Merriam-Webster English dictionary offers 14 distinct meanings for it. *Wikipedia* has 45 different definitions, including a movement for Democracy in Liberia, an alternative rock group from Australia, and a village in Poland. Even if we limit our attention to science, the definition remains elusive.

In recent years this question has drawn a lot of attention among philosophers and epistemologists, provoking an intense debate. A good summary of such a debate can be found in the *Stanford Encyclopedia of Philosophy*, in the article "Models in science" (Frigg and Hartmann, 2008). If we limit our attention to the semantic perspective, the *Stanford Encyclopedia of Philosophy* proposes that models have two distinct functions: a model can represent selected parts of the world, or it can represent a theory. On its own, this first distinction yields two very different concepts of the word "model."

If we now focus on the semantic definition of a model as a representation of selected parts of the world (the "target system;" hereinafter I shall use the term nature, which is more significant within the scope of this book), the *Stanford Encyclopedia of Philosophy* further distinguishes between models of phenomena and models of data. While every scientist heuristically understands this distinction, one could question this distinction by noticing that data are obtained through the observation of phenomena, and thus they can be considered as representations of phenomena in themselves. The debate becomes even more involved if we consider the problem from an ontological or epistemological point of view.

While the debate that the *Stanford Encyclopedia of Philosophy* outlines is of great interest and already provides some normative and interpretative concepts potentially useful to the scientific community, it is contained entirely within that collective construct of the human mind that we call philosophy. However, some aspects of the modeling process of vital importance for many empirical sciences lie in the interaction between the human mind and physical reality; these aspects, in the current philosophical debate around the definition of model are considered but not placed at the center of the attention.

In this section, I shall present a new theory where modeling is defined in the context of current theories of evolution. Such a perspective enables not only a unique definition of a model, but also an effective taxonomy of the different types of models that can be useful in many contexts.

Although I acknowledge that this theory is simply another point of view in the complex question of defining models in science from a philosophical perspective, and that it does not yield any radical innovation, rooted as it is in Popper's problem-solving theory (Popper, 1999), I claim that the particular grounding I use for my inference renders this viewpoint greatly interesting for most empirical sciences that are oriented toward problem-solving, such as biomedical engineering. But before that, it might be worthwhile introducing some fundamental concepts of the theory of falsification.

1.2.2   The theory of falsification

Popper (Popper, 1999) distinguishes between atomic statements, which, for example, are used to report a single observation, and universal statements, which aim to convey a general perspective, such as in a theory. Basic statements are atomic statements that have been confirmed by observation.

A universal statement can never be proved to be true; this is like saying that the probability associated with a universal statement is always zero. However, a universal statement can be proved to be false, if it is falsifiable. Thus, the first requirement for a scientific theory is that it must be formulated as a falsifiable universal statement.

Universal statements can be non-falsifiable; but if they are falsifiable this can be of different degrees. Indeed, the degree of falsifiability can be used as a metric to compare candidate theories. However, it should be noted that the degree of falsifiability of a universal statement says nothing about the falseness of the statement itself.

Once a theory has been chosen to explain a certain set of observations, we need to begin the falsification process; by challenging the theory with the most formidable attacks we shall be able to prove its "mettle," as Popper says. A theory will never be true, but the more extensive the list of falsification attempts that it has resisted, the more likely that it is true.

Given this, the scientific method becomes:

(a) Observe a phenomenon.
(b) Develop multiple theories that explain this observation.
(c) Express each of them with a falsifiable universal statement.
(d) Choose among them the one with the highest degree of falsifiability.
(e) Use the chosen theory to predict additional observations.
(f) If the theory is predictive, start to create falsification experiments.

If the theory is predictive, and none of the falsification experiments that we can devise is able to prove it false, we conclude that the theory is corroborated by those falsification experiments.

Popper's theories have been strongly debated, and particularly severe attacks came from scientists who accused Popper of not capturing in his theories the real process used in science to search for the truth. In spite of this, Popper's theories remain the best available attempt to provide scientists with some prescriptive arguments in support of their daily work.

In the following section, I shall rely heavily on the falsification theory to provide an operational definition of a predictive model.

1.2.3 What a model is, take one

The process that I propose is recursive. I shall first provide a descriptive definition based on certain characteristics that most of the literature agrees are specific to models. Using this first description, I shall speculate on the evolution of problem-solving skills in a hypothetical organism, showing that the development of modeling abilities is an imperative of such an evolutionary process. Lastly, I shall use this evolutionistic context to elaborate a new taxonomy for models.

A model is a cognitive construct, a product of the mind. This applies also to physical models, such as scale models; to create a scale model of a bridge, we must first decide which features have to be represented by the model, and how they are scaled down. In a way, we first create the model in our minds, and then build it.

A model is always finalized. Our mind can create multiple models of the same portion or reality, each to serve a specific purpose.

A model always has a representational function. In forming a model, we create a cognitive construct of finite complexity to represent a portion of nature of infinite complexity. This is possible using two idealization mechanisms: Aristotelian and Galilean (Frigg and Hartmann, 2008; Matthews, 2004). We use Aristotelian idealization when we recognize that only certain features are important, and neglect all the others. We use Galilean idealization when we build the model by making assumptions we know to be incorrect. Frictionless planes or massless objects are examples of Galilean idealization. In some cases a model also has a predictive function.

For completeness, it should be noted that I will reject alternative approaches where a model is presented as the model of a theory (Frigg and Hartmann, 2008), or where the aim of modeling is the "construction of a formal system for which reality is the only interpretation" (Carnap, 1958). While perfectly acceptable, these perspectives see the theory as an end, rather than a means, as more common in the context of problem solving.

Thus, we can summarize as follows:

A *model* is a finalized cognitive construct that provides an idealized representation of a portion of nature, and that can be used for descriptive or predictive purposes.

1.2.4 Modeling as an evolutionary imperative

1.2.4.1 All life is problem solving

This statement is the title of a collection of magisterial lectures that Karl Popper delivered toward the end of his career; it summarizes a good part of his life-long reasoning simply and clearly (Popper, 1999).

Popper argues that all biological entities constantly face problems, whose solution directly or indirectly affects their probability of surviving, reproducing, and thus transferring their particular genotype to the next generation. From this perspective the ability to solve problems is the essence of the evolutionary process. I want to extend this position, by arguing that modeling is the essence of the evolutionary process.

I shall do this with a simple thought experiment, where we follow the evolution of a hypothetical generic organism. In this exercise, I shall neglect every aspect related to reproduction and interaction with other organisms of the same species; the introduction of these factors would make the discussion much more complex but would not yield significantly different conclusions.

Let us start by assuming the existence of an organism so simple that it lacks any sensorial skill; this makes the organism totally unaware of its environment, and thus of the problems this poses. It can only behave randomly, hoping that this will save its life.

A first improvement is the introduction of sensorial mechanisms. With them the organism can now perceive certain features of the environment, and eventually also perceive as these features a change resulting from its actions. It can then use this sensorial feedback to explore, for each problem it faces, the entire space of the available solutions, searching for that which minimizes the adverse components of the sensorial feedback, or which at least makes it tolerable. This trial-and-error approach is the first and most basic problem-solving approach.

1.2.4.2   Remembering and comparing

The biggest drawback of this approach is that when faced again with the same problem, the organism would start again from scratch, and each time the probability of finding a good solution would be the same. So the next improvement we can imagine is the introduction of some memory. If the organism can remember which successful strategy was used last time it faced that problem, its survival chances increase dramatically. It should be noted that the term memory here does not necessarily indicate higher cognitive functions: a typical example would be an autonomic response to certain stimuli.

It is only when memory is considered that successful solutions to a certain problem can be considered expectations; according to Popper: "For the animal's behavior shows us that it expects that in a similar case the same ... [solution] will again solve the problem in question" (Popper, 1999). So each tentative solution that is found to solve a certain problem is an expectation, and thus a prediction, that next time we face this problem the same solution will work as well as it did this time.

Here is when the first emergence of the "model concept" occurs. The organism is capable of classifying problems on the basis of its sensorial input. Two situations that produce identical sensorial input can be considered the same problem, and thus solved using the same strategy. But what if the sensorial inputs are slightly different? In a monosensorial case, I could solve this by developing some ordinal relationship (e.g., based on intensity), but if I have complex multisensorial patterns, the problem becomes very difficult. The organism needs to develop some similarity criteria to compare sensorial patterns, so as to decide when a certain strategy, previously found successful to solve a certain problem (identified by a certain sensorial pattern) can also be used for another problem identified by a sensorial pattern similar to the first one. I would argue, also on the basis of modern data fusion theories, that the development of similarity criteria for complex sensorial patterns requires the creation of cognitive constructs that idealize the portion of nature under observation, which is exactly the definition I gave of "model."

Here it should be noted that a non-trivial by-product of such descriptive models is the ability to idealize problems, the first step toward generalization. The organism is now able to recognize a new problem as a member of a category of problems, all admitting the same solution.

Various philosophers have expressed this same view in different ways. Edgar Morin writes: "Knowledge cannot directly reflect reality, but only translate and reconstruct one reality into another reality" (Morin, 1993).

Another interesting argument, which is too long to be tackled here, is the role of descriptive modeling in the emergence of the two fundamental behaviors in intra-species interaction: imitation and communication. Both processes require as a first step that a certain idealization of the observations be made. They also require the sharing of these descriptive models between the interacting organisms, with interesting resonances with the concept of archetypes in psychology (Jung, 1981).

1.2.4.3 Prediction to overcome the limits of experience

Our organism is now capable of memorizing solutions to known problems, and recognizing similarity among problems. Still, this does not protect it from dangerous problems. We define a problem as dangerous when its direct observation seriously damages the organism, preventing a massive accumulation of experience on such a problem. To deal with dangerous problems, the organism has to develop a new skill: prediction. If the organism can recognize the sensorial patterns that anticipate the occurrence of a dangerous problem, it can undertake actions aimed at avoiding the problem or at reducing the damage it can produce.

The first approach in the development of a prediction is based on the assumptions of regularity and causation. If the dangerous problem the organism is observing has a causal relationship with some observable conditions, and if this relationship has sufficient regularity, after a sufficient number of observations the organism can build a predictive model by induction.

While this approach is extremely powerful, it requires that the organism expose itself to the dangerous problem more than once before a reliable inductive prediction can be made. And this cannot clearly be considered an ideal condition. We argue that this is the prime evolutionary motivation for the need of deductive prediction. The accuracy of deductive models does not depend on the number of observations; in theory, a single observation could be sufficient for our organism to develop a causation theory capable of predicting the emergence of that dangerous problem.

There is a third kind of logical reasoning, called *abduction*. First proposed by Charles S. Peirce, it describes the process of arriving at a possible explanation for an observed circumstance. To explain the difference between the three types of logic reasoning, we can use the classic example of the beans: if I take some beans from a bag, and they are all black, I can *induce* that all beans in the bag are black; if I know for sure that all the beans in that bag are black, when I take some beans from that bag I can *deduce* before observing that those beans will be black; if I know for sure that all the beans in that bag are black, and I see that some beans are black I can *abduce* that those beans come from that bag. Abduction provides another important capability to our organism: that of *guessing* a possible cause for an observed effect.

1.2.5   What a model is, take two

The hypothetical evolutionary process described above yields a new operational definition of a model: in the context of problem-solving, models are cognitive constructs providing idealized representations of nature that are used to generate tentative solutions to known problems. Although this definition has many operational advantages, it does not provide a taxonomic structure for the various types of model. Therefore, we need to develop some additional definitions.

The physical universe is characterized by a large number of observable states. In most cases a problem requires that certain observable states appear or that certain other states do not appear. We call the set of observable states of interest for our problem the "outputs," and the portion of universe that they characterize the "subject." Other observable states are also related to our outputs, in the sense that if they change, the outputs change. We call these other related states the "inputs," and the portion of universe that they characterize the "environment."

The definition of a cause–effect relation between the states of the environment and those of the subject is called "causation," and is a complex problem. Here I will only introduce operational definitions, recognizing that the identification of causation is a prime problem in science. First of all, let us accept the principle that the universe is entangled, and thus in principle any observable state characterizing a portion of the universe is somehow related to any other observable state; however, in many cases the sensitivity of one state to the changes of the other is negligible for all practical purposes. Thus, we can define two states as related if their causation is non-negligible with respect to our specific purpose. The entanglement implies that there might be relations not only between inputs and outputs, but also between inputs, within the portion of universe we call the environment. Thus, we define generic causation as the identification of all non-negligible relations between inputs and outputs; we define specific causation as the identification of the relations between the independent inputs and the outputs. With generic causation we know that changing certain inputs is sufficient to produce a change in the output, but we cannot tell which of those inputs is necessary to change the outputs; with specific causation we know which inputs are sufficient and necessary to change the outputs.

According to these definitions, the solution of a problem translates as "the identification of the inputs that produce the expected outputs during the interaction between the subject and the environment." Such knowledge will give us at least the ability to recognize in advance the appearance of the inputs that are associated with the expected outputs; in addition, if we are somehow able to control the inputs, we can produce the expected outputs.

There are three types of knowledge involved in this problem: knowledge of the expected outputs, knowledge of the inputs, and knowledge of the relation between the inputs and the outputs. The different types of predictive model can be categorized by the type of knowledge they incorporate.

- **Descriptive models** are models that incorporate only knowledge of the expected outputs, and provide a generic causation with a usually large set of inputs. They rely entirely and only on observation, do not attempt to establish a specific causation between inputs and the observed outputs, and do not speculate on why the outputs occur.
- **Inductive models** are models that incorporate knowledge of both the inputs and the expected outputs; they usually aim at specific causation. This is usually achieved through the introduction of systematic measurement of both inputs and outputs.
- **Deductive models** are models that incorporate knowledge of the inputs and the expected outputs, and also on the relation between them. They provide the best specific causation, and by describing the relationship between the inputs and the outputs using physical laws, they also provide a functional representation of the subject itself. In other words, deductive models contain knowledge not only of the inputs and outputs, but also on the subject itself.
- **Abductive models** are the most elusive to define; they are models that incorporate the ability to adjust the causation hypothesis as new observations are produced.

1.2.6 Conclusions

The most important result of this speculation is that – in the context of problem-solving – modeling appears to be an evolutionary necessity in the very first stages of the evolution of sentient organisms.

If correct, this line of reasoning could explain the difficulty in defining what a model is. Traditionally, modeling is seen as the result of higher cognitive functions, mostly associated with the formulation of theories; however, this is in part contradicted by the simplest representational functions that characterize every model. Our thought experiment suggests that, on the contrary, modeling might appear very early in the evolution of sentient organisms, and because of this it should be seen as a foundational mechanism of the organism–environment interaction.

These reflections carry some epistemological implications for empirical science. The outlook that I propose nullifies the distinction between observation and theorization, as it is typically made in empirical sciences in the distinction between experiment and model. I would argue that the simplest observation of nature cannot have any problem-solving usefulness if it is not already combined with some idealization process; what we call *descriptive modeling*. In this sense, any debate on the degree of truth that we can hope to achieve with the various approaches used in scientific investigation appears to be quite futile; from my argument it seems that we should judge our efforts of exploring nature only in terms of problem-solving efficacy. This is where we started from, and this is where we should always return.

1.3   The problem of validity

Every model is generated by starting from some idealizations, so how can we pretend that it is true? The problem of model validation is very complex, and in large part still debated. But first let me clarify that this is not simply a problem of computational models. A model is a cognitive artifact that we create in our minds; such a model can be physical, mathematical, procedural, etc. In biomedical research, there are human models, animal models, ex vivo models, in vitro models, and then there are mathematical models, which can be solved in closed form or numerically. In this sense, while an animal model has a physical reality that a mathematical model lacks, contrary to what many believe this does not, in itself, grant the animal model greater truth content than the mathematical model, as they are both affected by the idealizations we use to build them. So while I shall talk about numerical models, the problem of validity is much broader and involves any controlled condition we use to explore reality.

A numerical model can be true at two levels. Firstly, it can accurately predict the state of a portion of reality under certain conditions; secondly, it can provide insights on why that portion of reality assumes that state under those conditions. Not all models are capable of this second function: for example inductive models, in themselves, do not reveal anything on why something is happening, even if they can accurately predict it. In this sense, the problem of validation is related to the scope of a model.

If a model aims only to predict the state of a portion of reality under certain conditions, in as far as we can prove – by comparing the predictions of our models with quantifications of that state under controlled conditions (a controlled experiment) – that the model predicts reality with an error smaller than the required accuracy, we can conclude that the model is *validated*. Of course the validity of that model is limited to that scope, and to the set of conditions under which it is found to predict reality accurately. Only the regularity of physical phenomena provides a minimal justification for using the model to predict the state under different conditions; the further we go from the conditions of validation, the lower is the reliability of the model.

However, when a model aims to provide insight on why a certain portion of reality assumed a certain state under certain conditions, the whole concept of validation is meaningless. In that case, the best we can say is that the model *has not been falsified yet*, and neither has the theory it embodies, but we can never prove that it is true. The validity of a model with such an aim can only be accounted with respect to the number and extent of the attempts to falsify it that failed.

1.4 The role of predictive models in biomedicine

1.4.1 Biology vs. biomedicine

In the following, I shall repeatedly use the term "biomedicine" to indicate the whole corpus of knowledge developed by biological and medical sciences. The word is used to highlight the applied nature of this book: we are interested in the application of predictive multiscale models to the investigation of human physiology and pathology, with a clear goal toward the application of the results of this investigation to the practice of medicine. In this sense we are most interested in the part of biology that can help to improve the healthcare of human beings. As this focus will inevitably bias the positions we express in the following sections, it is fair to state this explicitly here.

1.4.2 The war against prediction

In spite of their popularity in other human activities, the introduction of predictive models in biomedicine has been somehow resisted, and their use is still almost negligible in most clinical practice. In part this is because in many cases reliable predictive models are a recent achievement. But it also partly arises from some cultural resistance, which has some serious roots in the epistemological debate around biology as a physical science.

Ernst Mayr is considered one of the leading evolutionary biologists of the past century; among other things Mayr's work has contributed to the revolution that led to the modern evolutionary synthesis of Mendelian genetics and Darwinian evolution, and to the concept of biological species. Thus, it is interesting to see the arguments Mayr uses against predictive models. Firstly, Mayr rejects the position that sees biology as a branch of physics (Mayr, 1996). Mayr rejects *vitalism* (the belief that the functions of a living organism arise from a vital principle distinct from physicochemical forces) and recognizes that all biological processes can be reduced to a sequence of events well described by physical and chemical theories, but argues that such theories alone cannot explain the observations of biology. His argument is long and articulate. However, it is important to notice that a key point in this argument is the role of downward causation, which prevents the researcher from understanding events at the organism level by studying only the molecular events. Because of this, biology is a separate domain of knowledge, knowledge that is separated and irreducible to that physics and chemistry. It is hard to contradict this position; however, some derivations are of some concern in the context of this book.

In his 1998 book (Mayr, 1998), Mayr argues against Popper's falsification theory because, "In probabilistic theories a single falsification is not sufficient to reject a theory." Similar arguments are used against the use of mathematical models, which are accused of oversimplifying the complexity of nature, and completely overlook the probabilistic nature of many of them. These views, shared by many biomedical researchers, tend to ignore the fact that modern mathematics has very powerful methods to represent and manipulate probabilistic phenomena. But in my opinion there is nothing in Popper's theories that imposes determinism as a requirement for the falsification theory. On the contrary, it is mandatory to account for the so-called falsification asymmetry when we design statistical tests; this is done by posing the research question in a way that puts the weight of the inter-subject variance on the side opposite to our question, so that the power of our statistical test works in favor of the confirmation, or, which is the same, the variance works in favor of its falsification.

It is clear that in many cases the complexity of the problems we encounter in biomedicine, and the limitations of our knowledge, force us to come to severe compromises with the methodological rules of science. In applied sciences such, as medicine or engineering, this is not perceived as a terrible sin. However, it is important to recognize when such compromises are made, and to admit that whenever possible they should be avoided. So it is true, as Mayr says, that, "In evolutionistic biology to explain some observations you have to elaborate descriptions as historical tale," and that this makes it very difficult to falsify a theory. However, this should not be taken to mean that Popper is wrong, but rather that such evolutionistic historical tales are probably the best we can produce as yet, but their potential for scientific truth is somehow limited, precisely because of their limited falsifiability.

Another argument that Mayr proposes is the impossibility of using predictive models in biology. In his otherwise illuminating commentary on the debate between Mayr and Popper, Tom Settle wrote: "Biology repeatedly testifies to the impossibility of forecast, even of rough forecast. For example, no one can forecast either the pace or the direction of evolution" (Settle, 1989). Settle is clever in choosing his example. Evolution is the emergence produced by complex (non-linear and possibly chaotic) interactions between the individual, the species, and the environment; as such, it is probably the most challenging aspect of biomedicine to be modeled and predicted. In addition, the typical time frame of evolution renders impractical any attempt of designing a falsification experiment in a classic way. But I claim that this is very difficult, but not utterly impossible. And evolution is an extreme case. There are many other phenomena in biomedicine that pose much smaller challenges. A key problem, which this book tries to address, the prediction of the maximum load that the skeleton of a given subject can withstand without fracturing – a problem that has significant clinical implications – can unquestionably be forecast.

Thus, we may conclude that, at least in principle, the theories we use to describe the physiology and the pathology of biological organisms are falsifiable, and that it is possible to develop predictive models from these theories, although in many practical cases this is so difficult as to appear impossible. The methods that I propose in this book are an attempt to solve this methodological impasse and to bring falsification theory and predictive models to mainstream biomedical research.

1.4.3 Why biomedicine needs prediction

1.4.3.1 Predictions in problem-solving activities

The fact that predictive models of biological systems are possible does not necessarily imply that they are useful. But they are. I shall follow a formal argument, which relates the role of predictive models to the application of the scientific method. As this reasoning strongly relies on Popper's falsification theory, we see that the problems of falsification and prediction are entangled.

But before I begin such speculative discussion, I would like to stress the other, more empirical role of prediction. As I explained previously, problem solving is the most natural activity for all living creatures. The ability to forecast how a certain possible action would affect the problem we are facing is a dramatic competitive advantage, as it significantly increases our chance of survival. Thus, human beings have made predictions as far back as we can remember, as a species.

The daily practice of medicine is, for the most part, a forecasting exercise. In some cases, this prediction is obtained through deductive reasoning, and inductive modeling. In many others, it is left somehow implicit within the metaphysics gray zone, which is always present in the practice of medicine. But whether this happens explicitly or implicitly, medicine is first and foremost the ability to predict, "What will happen if...?" In this context, it is not surprising if some practitioners view the introduction of predictive models with some hostility. Their adoption forces the explicit formulation of all hypotheses, assumptions, and observations, and produces a constraining prediction, frequently quantitative, which – as we shall see in the following section – is easier to prove false than other types of statement. Surprisingly, this is very close to the manifesto of so-called "evidence-based medicine" (Sackett, 1996), a methodology that aims to steer medicine toward decisions based on explicit and documentable processes. Again, criticisms of the introduction of predictive models in clinical practice seem more justified by practical difficulties than by any substantial issue. However, if one accepts the challenge that the use of predictive models in the clinical practice involves, the potential advantages are significant.

1.4.3.2 Prediction forms the basis of the scientific method

Even the most severe critics of the so-called *physicism* in biology, such as Mayr, accept that the scientific method has proven superior to any other method of investigating nature. But even if we accept the scientific method, this does not automatically imply that we should accept predictive models. Specifically, where do predictive models stand with respect to the scientific method? The aim of this section is to advocate that predictive models are the foundation of the so-called scientific method and that, among the many methods we have to describe a theory; they are far superior from an epistemological point of view.

I now want to substantiate the statement that among the many ways we have to express a tentative solution (e.g., a theory) to a given problem, a predictive model is to be preferred. This is related to the difference between the primitive trial-and-error pattern I described above, and the scientific method. The fundamental improvement of the scientific method over trial-and-error problem solving is that to find the best solution to a problem we do not need to try it. By combining the solutions we have found so far to related problems (our experiential knowledge) we can develop a logical framework, usually of a deductive nature, that makes it possible to know which solution is the most suitable for a given problem without needing to test all of them. Again with Popper: "The distinctive feature of science is conscious application of the critical method." Science enables the formulation and the presentation of each tentative solution in some language, so that "They become objects of consciously critical investigation." So in science we can express and critically revise each tentative solution without the need to put it to the test in practice.

The essence of Popper's theory is that the critical method should involve the evaluation of each tentative solution in the most unfavorable conditions; in other words, we should try as hard as we can to prove that the tentative solution does not work; if we fail to prove it wrong, we can assume that, so far, that solution is the best among those considered. Popper calls this process "falsification." In this context, the language we use to express our tentative solution is vital. In all cases, we should prefer a representation of the tentative solutions that makes it as easy as possible to prove the models' eventual falsity; this is called *falsifiability*. It is easy to see that among all those possible ways to express a tentative solution, a predictive model is the easiest to falsify. So far as the problem we deal with admits some experimental measurability, we simply have to predict those quantities with our models, and then compare the predictions with the measurements. The farther the prediction is from the measurement, the stronger is the evidence of its falsity. Because of this, we can conclude that a predictive model is the best way to express a theory, as it is maximally falsifiable.

1.5   Integrative research and multiscale modeling

1.5.1   Complexity comes from many places

In the previous section, I reported various criticisms of the application of predictive models to the investigation of research problems in biomedicine. Most of these arguments revolve around the difficulty of approaching, with the classic methods of physics and chemistry, the study of systems that exhibit regularities only in a statistical sense. Another problem frequently pointed out is that, for many biological systems, the causal relationship between the evolution of the system in time and some of its initial conditions is known, but any attempt to define a quantitative correlation between the two fails, as very small changes in these initial conditions can lead to completely different results in the evolution of the system. Although I questioned the belief that these difficulties lead to the methodological impossibility of using predictive models in biomedical research, it is undisputable that such difficulties do exist, and that their existence makes the modeling of biological systems overwhelming challenging.

The great variability of the observations made on biological systems, once we have ensured that our measurement methods do not add significant variability because of insufficient repeatability, can be distinguished as being of two types: inter-subject and intra-subject. As we shall see in the following chapters, the inter-subject variability can be addressed by using subject-specific models, i.e. models that are identified with parameters directly measured on the subject of interest, and not taken from average values of a population. Intra-subject variance is more complex to address, because it can have different origins.

The first case is when we do not apply the reductionist approach effectively. The stress and strain induced in bones by daily activities can vary considerably; however, this is mostly because of the variability of the physical activities a person undertakes in his or her daily life, and the variability of the neuromotor control involved. But if we limit our observation to the stress and strains induced in a given bone by a given musculoarticular loading, then the variance is very small. By choosing our subject of investigation wisely, we can sometimes remove a lot of intra-subject variance.

A second case is when the observed variance is a function of the biological system, e.g. for redundancy reasons. Let me give an example to clarify this concept. While we move, at every instant the resultant of the external forces and moments that act on our body (inertia, the interaction with the ground, etc.) must be in instantaneous equilibrium with the forces and moments generated by the muscles and transmitted by the joints and ligaments. A classic problem in biomechanics is the determination of the force expressed by each muscle bundle in a certain instant of a particular movement. This problem is difficult because in most cases the number of muscle bundles is much greater than the number of degrees of freedom of the kinematics chain that represents the articulated skeleton. Thus, there is an infinite number of activation patterns through which the nervous system can excite the muscles to ensure instantaneous equilibrium. A classic strategy in biomechanics is to postulate that the neuromotor control activates the muscle bundles with a strategy that aims to minimize one or more factors: energy expenditure, muscle stress, fatigue, etc. By minimizing this cost function, we obtain a solution to this problem, which would otherwise be undetermined. Unfortunately, this approach is effective only in a few cases, and this should not surprise us. If everything is fine, one probably walks to minimize the energy expenditure, and to avoid muscle damage. But if that subject is being chased by a tiger, it is reasonable to expect that the neuromotor strategy will change toward maximizing the performance! If the subject had a small trauma to a leg muscle the day before, the neuromotor strategy will be adapted to protect that painful muscle as much as possible. It is possible to make an infinitely long list of such cases. The point is that the neuromotor system controls muscles in many different ways; if we pretend that the control strategy is fixed, we commit what in some cases is an unacceptable error, which appears in our measurements as intra-subject variance.

If we now look closely at the second problem mentioned at the beginning of this section, the difficulty of defining a quantitative causal relationship between the initial state and the temporal evolution of certain biological systems, again we shall be confronted with different cases.

In some cases, the problem is that what we observe in reality is not a direct cause–effect relationship, but rather an indirect relationship produced by the strong systemic *entanglement* typical of organisms.

In other cases the causation relationship exists, and it is reasonably deterministic, but it has a strong non-linear behavior. What do we mean by the term "non-linear?" A linear system is a system that satisfies the *superposition principle*:

The net response at a given place and time caused by two or more stimuli is the sum of the responses that would have been caused by each stimulus individually.

In non-linear systems this principle does not apply.

Another way of looking at it is that in a non-linear system, where the quantity A depends on the quantity B, there are, within the same system, conditions where small changes of B produce small changes of A, and other conditions where such small changes of B produce very large changes of A. Sometimes the sensitivity to the initial conditions is so high that at first glance the behavior appears totally random; this is what mathematicians call a *chaotic system*.

Sometimes these two problems combine, and we end up with a system where we can establish cause–effect relationships only on a statistical basis, owing to the inherent variability of the observed phenomena, and these relationships are strongly non-linear.

However, if we begin to look more closely at all these problematic cases, we start to notice a pattern. In some cases, the problem derives from the fact that the sub-system we are observing is not isolated from the other sub-systems, as we optimistically speculated. So in the example above about muscles, the musculoskeletal sub-system is not isolated by the neuro-cognitive sub-system. An example: if I am depressed, my walking posture is different, and this also affects the musculoskeletal biomechanics. A recent study showed that the distance an older person can walk in six minutes is affected by a large number of physiological and psychological factors, and emotional state is not a marginal one (Lord and Menz, 2002).

In other cases, the problems are caused by strong systemic entanglement, as mentioned. Figure 1.1 shows the integrated cellular network of *Saccharomyces cerevisiae*, a type of budding yeast.

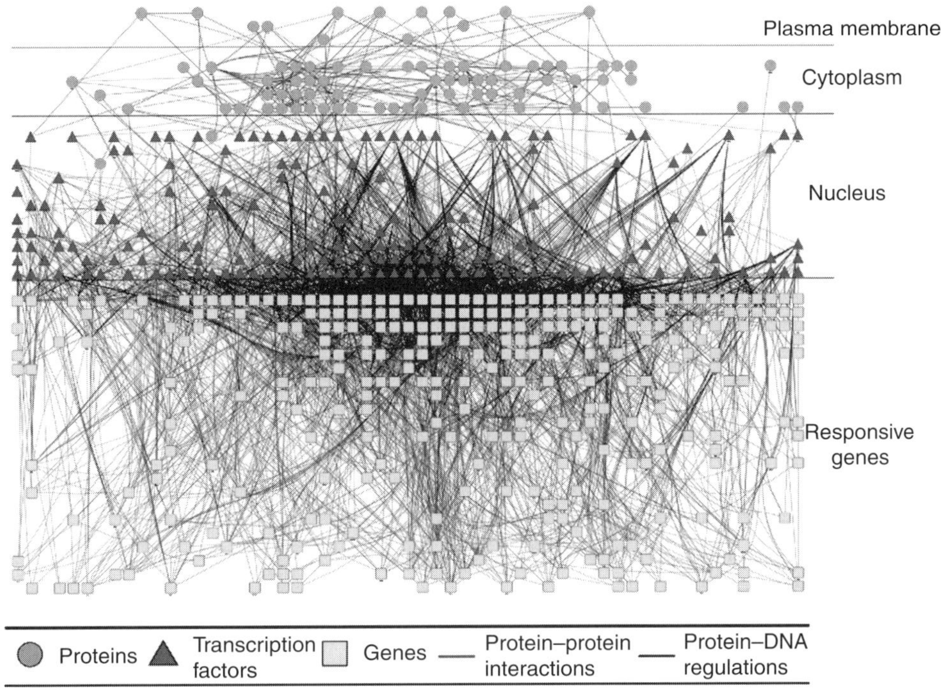

Figure 1.1 The *Saccharomyces cerevisiae* integrated cellular network under hyperosmotic stress. The network consists of 309 genes, 162 transcription factors and 82 proteins, with 2836 protein–DNA regulations and 301 protein–protein interactions. Reproduced from (Wang and Chen, 2010) under the terms of the Creative Commons Attribution License, © 2010 Wang and Chen; licensee BioMed Central Ltd.

When so many intricate interdependencies exist between the parts forming a biological sub-system, the observation of *emergences* is not unlikely, i.e. the appearance of the complex patterns from a multiplicity of relatively simple systemic interactions. In a way, the concept of emergence is already implied in the famous Aristotelian statement, "The totality is not, as it were, a mere heap, but the whole is something besides the parts." (Aristotle, 350 BC).

Last, but not least, in many cases the difficulties described previously arise from the interaction of processes taking place at very different temporal or spatial scales. A very simple example: if you remove the anterior cruciate ligament from a sheep's knee, and subject it to a tensile rupture test, the force–displacement curve will look like the one depicted in Figure 1.2.

A ligament looks like a rope, a strip of soft material. Every engineer would expect that such an apparently simple structure, when subjected to a tensile test, would respond with a linear force–displacement curve, at least until the stresses start to exceed some proportionality limit. Instead, what we see is strong non-linear behavior at low loads and quite linear behavior afterward, until rupture. This unexpected behavior can be explained if one takes into account the multiscale

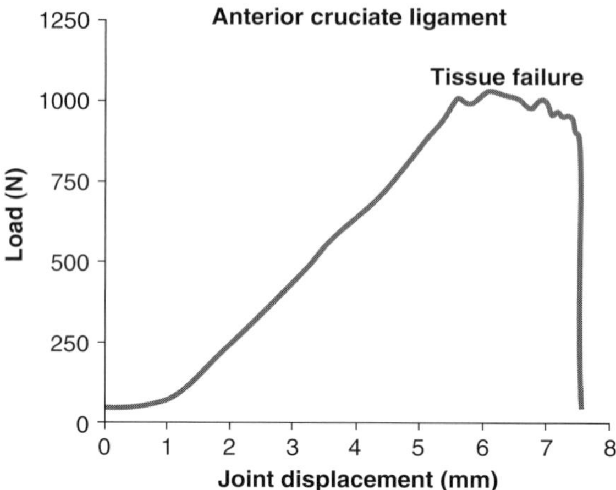

Figure 1.2   Force–displacement curve recorded during a tensile rupture test on a bone–ligament–bone complex of a sheep's anterior cruciate ligament.

morphology of ligaments. If we observe the ligament tissue at rest with a scanning electron microscope, we see that the structure is coiled. When a small load is applied, the tissue fibrils first uncoil (producing the characteristic low-stiffness foot in the curve in Figure 1.2) and then stretch according to a perfectly linear law until damage starts to appear. This example shows how complex non-linear behaviors are frequently produced by complex interactions between processes taking place at very different dimensional or temporal scales.

1.5.2   Modeling biological systems

From the previous section, we know that biological systems frequently behave in ways that are difficult to model for a number of reasons. However, all these reasons have something in common.

If we repeatedly subject a biological system to a series of stimuli and observe the organism's response, in most cases we shall notice that the response to the same stimuli varies considerably across repetitions, and that frequently the cause–effect relationship between the stimuli and the response is strongly non-linear. In some cases, we can solve the problem by focusing our attention to a limited portion of the biological system, small enough to exclude most systemic interactions; the behavior of the system becomes tractable, and it is possible to build an accurate predictive model.

But in many other cases, this reduction is not possible. When the response we are interested in results from the interaction between sub-systems, or when the process we want to study cannot be separated entirely from other biological processes, or when the complexity we observe emerges from an interaction across scales, the reductionist approach will inevitably fail. As the old saying goes, "If you can't beat them, join them"; if we cannot get rid of the systemic nature of the process, we should find a way to embrace it. And this brings me to the main topic of this book: how to model a biological system (namely the skeleton), to account for its systemic and multiscale nature.

1.5.3 Structure of the book

In this book, I shall try to answer this complex question, at least partly. The book is organized in eight chapters. After this introduction, a chapter is devoted to the methodological issues that we need to consider before we can start developing multiscale models of the skeleton. Then we shall tackle the problem, first working separately at each relevant dimension: we shall spend a whole chapter on skeletal modeling at each of the body, organ, tissue, and cell levels. In Chapter 7 I shall finally start putting together the pieces, and create multiscale models to be used in some relevant clinical applications. In Chapter 8 I shall look forward, envisioning a set of methods and technologies that, once established, will make the development of integrative models a simpler process than it is today.

2 Methodological aspects

A brief description of some computational mechanics methods that are used throughout the rest of the book.

2.1 Introduction

While this book does not aim at teaching computational biomechanics, in this chapter I shall review a few concepts of mechanics and computational mechanics that will provide a methodological foundation for the following chapters, where instead I shall focus on the biomedical application.

2.2 Elements of dynamics of rigid articulated systems

To investigate human motion we can assume that the skeleton is made of rigid elements (bones) connected by joints to form an *articulated system*. By considering the forces generated by the muscles we can then predict the kinematics of the skeleton (forward dynamics), while by considering the kinematics we can compute the joint moments, and from those derive the muscle forces (inverse dynamics). This section aims to provide some theoretical basis for the dynamics of articulated systems.

2.2.1 Single-body dynamics

Newton's second law, for a material point of mass m, position x, velocity $\dot{x}$, and acceleration $\ddot{x}$, subject to an external force F can be written as:

$$m\ddot{x} = F. \tag{2.1}$$

When elastic (conversative) and viscous (dissipative) reactions are present, this equation becomes:

$$m\ddot{x} - c\dot{x} - kx = F. \tag{2.2}$$

Generalizing the result for the material point to bodies with more complex three-dimensional geometry and mass distribution (a continuous rigid body), the pose (position and orientation) of the body is defined by the vector $[x\ y\ z\ \theta_x\ \theta_y\ \theta_z]^{\mathrm{T}}$. Now the equations of motion must also account for the rotational equilibrium. The

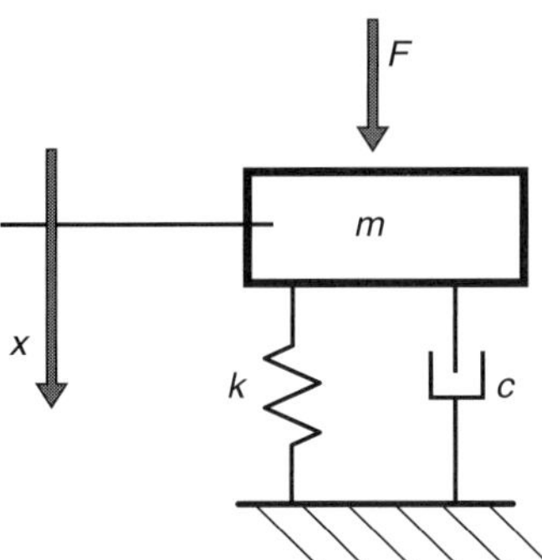

Figure 2.1   Spring, mass, and damper system.

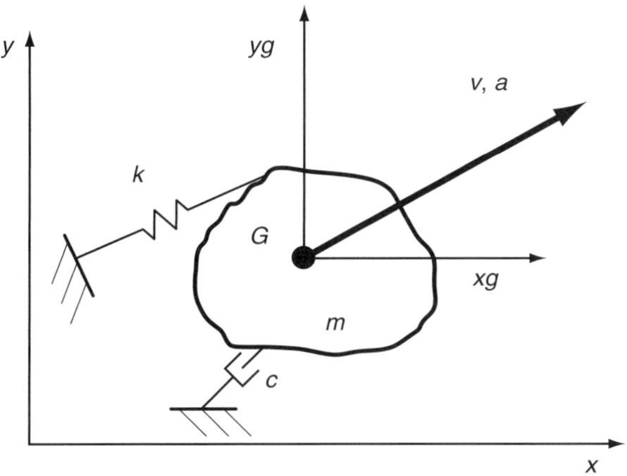

Figure 2.2   Two-dimensional spring–mass–damper system, where the body of mass m, center of gravity G, has a generic shape.

resulting moment produced by all inertial actions in a rigid body with respect to the reference system 0 is:

$$M_0 = -J_0\omega' - \omega^{\mathrm{T}}J_0\omega. \tag{2.3}$$

The inertial properties of a body, with respect to a generic $<x, y, z>$ reference system, are expressed by the tensor of inertia as:

$$J_0 = \begin{bmatrix} \int_m (x^2 + y^2)\,\mathrm{d}m & -\int_m xy\,\mathrm{d}m & -\int_m xz\,\mathrm{d}m \\ -\int_m xy\,\mathrm{d}m & \int_m (x^2 + y^2)\,\mathrm{d}m & -\int_m yz\,\mathrm{d}m \\ -\int_m xz\,\mathrm{d}m & -\int_m yz\,\mathrm{d}m & \int_m (x^2 + y^2)\,\mathrm{d}m \end{bmatrix} = \begin{bmatrix} J_x & -J_{xy} & -J_{xz} \\ -J_{xy} & J_y & -J_{yz} \\ -J_{xz} & -J_{yz} & J_z \end{bmatrix}. \tag{2.4}$$

It is possible to demonstrate that there is always a reference system for which the inertia tensor becomes diagonal:

$$J_G = \begin{bmatrix} J_{xx} & 0 & 0 \\ 0 & J_{yy} & 0 \\ 0 & 0 & J_{zz} \end{bmatrix}. \tag{2.5}$$

The diagonal elements J_{xx}, J_{yy}, J_{zz}, called the *principal moments of inertia*, represent the eigenvalues of the matrix J_0.

The eigenvectors of the matrix are called the *principal axes of inertia*. The moments of inertia of the body with respect to a generic axis passing through the center of mass of the body, can be expressed as the distance from the center at which such an axis intersects an ellipsoid whose semi-axes, oriented as the principal axes, have length J_{xx}, J_{yy}, and J_{zz}. Such an ellipsoid is called an *ellipsoid of inertia*.

2.2.2 Articulated systems

We define the *degrees of freedom* as the number of values that must be assigned to univocally determine the spatial configuration of a rigid system. If the system is formed by a single element (rigid body), this has six degrees of freedom; three that define the position, and three that define the orientation. For an *articulated system*, formed of n elements connected to each other by articulated joints, the number of degrees of freedom is equal to $6n$ less the number of degrees of freedom suppressed by the joints:

$$n_{gdl} = 6n - 5C_1 - 4C_2 - 3C_3 - 2C_4 - C, \tag{2.6}$$

where C_i indicates the number of joints in the system that leave i degrees of freedom unconstrained.

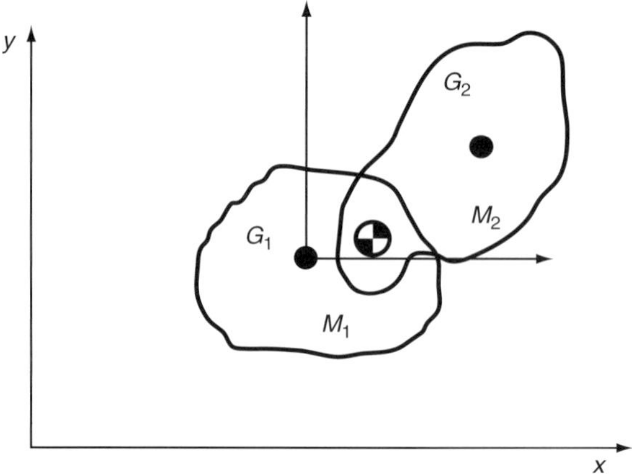

Figure 2.3 Schematic of a two-segment articulated system.

2.2.3 Lagrange equations

The kinetic energy of a rigid body in motion can be expressed as:

$$E = \frac{1}{2} m v_G^2 + \frac{1}{2} \omega^{\mathrm{T}} J_G \, \omega, \tag{2.7}$$

where G is the center of mass and J_G is the polar moment of inertia.

There are many methods for investigating the dynamics of an articulated system; one quite popular method uses Lagrange equations. Under some assumption on the system's joints the position of a generic point P can be expressed as a function of n *generalized coordinates* (and eventually of time t):

$$P = P(q_1, q_2, q_3, \ldots, q_n, t). \tag{2.8}$$

A virtual displacement (infinitesimal and compatible with the constraints) is given by:

$$\partial P = \sum_{k=1}^{n} \frac{\partial P}{\partial q_k} \partial q_k. \tag{2.9}$$

By setting dynamic equilibrium of the forces we obtain the Lagrange equations:

$$\frac{\mathrm{d}}{\mathrm{d}t}\left(\frac{\partial E}{\partial \dot{q}_k} \right) - \frac{\partial E}{\partial q_k} + \frac{\partial U}{\partial q_k} = Q_k, \tag{2.10}$$

where E is the kinetic energy, U is the potential energy, Q_k is the generic generalized non-conservative force, and q_k are the generalized coordinates of the system.

2.2.4 Numerical solution of articulated system dynamics

In most cases, the equations of motion of an articulated system are given as a second-order system of *ordinary differential equations* (ODE) in the form:

$$M(q_k)\ddot{q}_k = f(q_k, \ddot{q}_k, t). \tag{2.11}$$

This form can be transformed to a first-order system:

$$M(q)\ddot{q} = f(q, \ddot{q}, t) \Rightarrow \begin{cases} \dot{q} = v \\ M(q)\dot{v} = f(q, v, t) \end{cases}. \tag{2.12}$$

Such systems can be solved numerically using many methods, which in general ensure a rapid convergence toward the solution. Here, I shall briefly discuss only the two most popular, backward Euler and fourth-order Runge–Kutta.

The basic Euler scheme replaces the derivative with a finite difference:

$$\dot{y}(t) \approx \frac{y(t+h)-y(t)}{h}$$
$$y(t+h) \approx y(t)+h\dot{y}(t)$$
$$y(t+h) \approx y(t)+hf(t,y(t))$$

(2.13)

This formula is used iteratively to calculate progressively improved approximations of $y(t)$. Given a time step h such that $t_1 = t_0 + h$; $t_2 = t_0 + 2h$; etc., the initial condition $y(t_0) = y_0$, and a numerical estimate y_n of the value $y(t_n)$, we can compute the estimate at the next time step as:

$$y_{n+1} = y_n = hf(t_n, y_n).$$

(2.14)

This method, also known as *forward Euler*, tends to accumulate errors with increasing time step. It is thus preferable to use a different approach called *backward Euler*, where the derivative is approximated as:

$$\dot{y}(t) \approx \frac{y(t)-y(t-h)}{h},$$

(2.15)

which produces the following iterative scheme:

$$y_{n+1} = y_n = hf(t_{n+1}, y_{n+1}).$$

(2.16)

This means that to compute y_{n+1}, we need to solve another iterative cycle, for example using a Newton–Raphson scheme (see Section 2.3.9):

$$\Delta y_{n+1}^i = y_{n+1}^{i+1} - (y_n + hf(t_{n+1}, y_{n+1}^{i-1}))$$
$$y_{n+1}^i = y_{n+1}^{i-1} + \Delta y_{n+1}^i$$

(2.17)

Another approach is to use more points in the interval $[t_n, t_{n+1}]$. The most popular of these methods is the *fourth-order Runge–Kutta* (RK4) method:

$$y_{n+1} \approx y_n + \tfrac{1}{6}h(k_1 + 2k_2 + 2k_3 + k_4),$$
$$t_{n+1} = t_n + h$$

(2.18)

where y_{n+1} is the RK4 approximation of $y(t_{n+1})$ and:

$$k_1 = f(t_n, y_n)$$
$$k_2 = f(t_n + \tfrac{1}{2}h, y_n + \tfrac{1}{2}hk_1)$$
$$k_3 = f(t_n + \tfrac{1}{2}h, y_n + \tfrac{1}{2}hk_2)$$
$$k_4 = f(t_n + \tfrac{1}{2}h, y_n + \tfrac{1}{2}hk_3)$$

(2.19)

These are only two of the plethora of numerical methods used to solve ODE problems. However, in most cases where stability is not an issue, the only difference is efficiency, which is not a major problem, as ODE models are computationally much less demanding than partial differential equation problems, generally speaking.

However, there are situations where the differential equations that describe our problem cannot be written in ODE form. A typical example is when the mechanism contains closed loops (i.e., in the simulation of locomotion when both feet are on the ground). This adds some algebraic equations to the system, and the resulting mathematical form is called a *differential algebraic equation* (DAE). However, a discussion of the numerical methods used to solve such problems is beyond the scope of this book. Those interested in expanding the argument can find a complete discussion in Eich-Soellner and Führer (1998). Another interesting book on the subject is Ascher and Petzold (1998).

2.3 Brief notes on computational solid mechanics

When studying the biomechanics of the skeletal system at the organ and tissue levels, I shall describe bones as deformable solids, whose behavior can be modeled using the theory of *solid mechanics*, and a numerical method called the *finite-element method*. This section provides some essential theoretical basis on these topics.

2.3.1 Elements of tensor calculus

2.3.1.1 Notation

I shall use the *Einstein summation convention*, which states that when in an equation an index is repeated twice in a single term, this implies a summation over all possible values of that index. So for example the equation of a plane:

$$a_1{}^1x + a_2{}^2x + a_3{}^3x = p \tag{2.20}$$

can be written as:

$$\sum_{i=1}^{3} a_i{}^i x = p \tag{2.21}$$

but under the summation convention it will be written as:

$$a_i{}^i x = p. \tag{2.22}$$

Another useful notation is the Kronecker delta:

$$\delta_{ij} = \begin{cases} 1 \rightarrow i = j \\ 0 \rightarrow i \neq j \end{cases}. \tag{2.23}$$

2.3.1.2 Transformation of coordinates

The change of reference system, or transformation of coordinates, is a central point of tensor calculus. The triplet of independent variables x_1, x_2, x_3 can be considered as the triplet of coordinates that define the position of a point with respect to a reference system. The equation

$$\bar{x}_i = f_i(x_1, x_2, x_3), \qquad i = 1, 2, 3 \tag{2.24}$$

defines a transformation of coordinates, while the equation

$$x_i = g_i(\bar{x}_1, \bar{x}_2, \bar{x}_3), \qquad i = 1, 2, 3 \tag{2.25}$$

defines its inverse transformation. For the purposes of this book, we shall focus on transformations that are *admissible and proper* over a region Ω, which means that they are represented by single-value functions that are continuous and differentiable over Ω, with the determinant of the Jacobian matrix positive for all Ω:

$$J = \left| \frac{\partial \bar{x}_i}{\partial x_j} \right| = \begin{vmatrix} \dfrac{\partial \bar{x}_1}{\partial x_1} & \dfrac{\partial \bar{x}_1}{\partial x_2} & \dfrac{\partial \bar{x}_1}{\partial x_3} \\[2ex] \dfrac{\partial \bar{x}_2}{\partial x_1} & \dfrac{\partial \bar{x}_2}{\partial x_2} & \dfrac{\partial \bar{x}_2}{\partial x_3} \\[2ex] \dfrac{\partial \bar{x}_3}{\partial x_1} & \dfrac{\partial \bar{x}_3}{\partial x_2} & \dfrac{\partial \bar{x}_3}{\partial x_3} \end{vmatrix} > 0 \qquad \text{in } \Omega. \tag{2.26}$$

2.3.1.3   Metrics of Euclidean spaces

A metric space is a mathematical construct made of two elements: a set X and a distance function d, which associate two points, x and y, of X to a real number $d(x, y)$, which we shall call the "distance between x and y," and which fulfil three conditions:

- Every distance between two points is positive or zero, being zero only when the two points are coincident;
- The function is symmetric: the distance of point x from point y is equal to that of point y from point x;
- Given three points x, y, and z, the distance $d(x, y)$ cannot be greater than $d(x, z) + d(z, y)$.

The Euclidean space is a special type of metric space in real coordinates on which are defined a scalar product, a vector norm, and geometry based on Pythagoras's theorem, which is indeed called *Euclidean metrics*.

To define the reference system, we need the definition of length in that reference system. Let us consider a Euclidean three-dimensional space within which is defined an admissible and proper transformation from a Cartesian reference system to a generic one:

$$\theta_i = \theta_i(x_1, x_2, x_3). \tag{2.27}$$

We assume the existence of the inverse transformation

$$x_i = x_i(\theta_1, \theta_2, \theta_3), \tag{2.28}$$

so that it is possible to establish a bi-univocal correspondence between the two sets of coordinates. Let us define the components of a segment through the differentials

dx_1, dx_2, dx_3. Since the coordinates x_i are defined in the Cartesian reference system, the length of the segment can be expressed as (using the summation convention)

$$(ds)^2 = dx^i \, dx^i. \tag{2.29}$$

If we apply the transformation defined in Eq. (2.28), we can write

$$dx^i = \frac{\partial x_i}{\partial \theta_k} \, d\theta^k. \tag{2.30}$$

Substituting Eq. (2.30) into Eq. (2.29) we obtain

$$(ds)^2 = \frac{\partial x_i}{\partial \theta_k} \frac{\partial x_i}{\partial \theta_m} \, d\theta^k \, d\theta^m. \tag{2.31}$$

If we define

$$g_{km}(\theta_1, \theta_2, \theta_3) = \frac{\partial x_i}{\partial \theta_k} \frac{\partial x_i}{\partial \theta_m}, \tag{2.32}$$

we can write

$$(ds)^2 = g_{km} d\theta^k \, d\theta^m, \tag{2.33}$$

where the *functions* g_{km} are symmetric in k and m.

We can now extend the concept of the length of a segment of a generic reference system:

$$\theta_i = \theta_i(\bar{\theta}_1, \bar{\theta}_2, \bar{\theta}_3), \tag{2.34}$$

$$d\theta^k = \frac{\partial \theta_k}{\partial \bar{\theta}_l} \, d\bar{\theta}^l, \tag{2.35}$$

$$(ds)^2 = g_{km} \frac{\partial \theta_k}{\partial \bar{\theta}_l} \frac{\partial \theta_m}{\partial \bar{\theta}_n} \, d\bar{\theta}^l \, d\bar{\theta}^n, \tag{2.36}$$

$$g_{ln}(\bar{\theta}_1, \bar{\theta}_2, \bar{\theta}_3) = g_{km}(\theta_1, \theta_2, \theta_3) \frac{\partial \theta_k}{\partial \bar{\theta}_l} \frac{\partial \theta_m}{\partial \bar{\theta}_n}, \tag{2.37}$$

$$(ds)^2 = g_{ln} \, d\bar{\theta}^l \, d\bar{\theta}^n. \tag{2.38}$$

The functions g_{ln} are the metric functions in the reference system $(\bar{\theta}_1, \bar{\theta}_2, \bar{\theta}_3)$. Equation (2.37) is called a *transformation law*, and defines the relationship between the metric functions of two generic reference systems: this transformation is admissible and proper.

2.3.1.4   Definition of a tensor

The transformation law, which defines how the elements of a quantity change with a transformation of coordinates, characterizes the nature of that quantity. In classical mechanics, mass is an invariant with respect to the reference system, whereas the velocity vector changes if we change reference system. A quantity is called a *tensor* if and only if its transformation law fulfils certain criteria.

We define a quantity $\bar{t}^{\,\alpha_1\cdots\alpha_p}_{\beta_1\cdots\beta_q}$ *tensor field of rank* $r = p + q$, *contravariant of rank p and covariant of rank q*, if its components transform between two generic reference systems according to the equation:

$$\bar{t}^{\,\alpha_1\cdots\alpha_p}_{\beta_1\cdots\beta_q} = \frac{\partial\bar{\theta}^{\alpha_1}}{\partial\theta^{k_1}}\cdots\frac{\partial\bar{\theta}^{\alpha_p}}{\partial\theta^{k_p}}\cdot\frac{\partial\theta^{m_1}}{\partial\bar{\theta}^{\beta_1}}\cdots\frac{\partial\theta^{m_q}}{\partial\bar{\theta}^{\beta_q}}\,t^{k_1\cdots k_p}_{m_1\cdots m_q}. \tag{2.39}$$

If we limit our attention to Cartesian reference systems, this expression becomes much simpler:

$$\frac{\partial x_k}{\partial\bar{x}_i} = \frac{\partial\bar{x}_i}{\partial x_k}. \tag{2.40}$$

In other words, if we transform between Cartesian systems the distinction between contravariance and covariance disappears. In the following, we shall be interested only in tensors of rank 0, 1, and 2. A quantity is called a *scalar field* (a tensor field of rank zero) if it has a single component, and this does not change with the changing of reference system:

$$\bar{\phi}(\bar{x}_1,\bar{x}_2,\bar{x}_3) = \phi(x_1,x_2,x_3). \tag{2.41}$$

A quantity is called a *vector field* (a tensor field of rank one) if it has three components that transform at the change of the reference system according to the law:

$$\bar{\eta}_i(\bar{x}_1,\bar{x}_2,\bar{x}_3) = \frac{\partial x_k}{\partial\bar{x}_i}\,\eta_k(x_1,x_2,x_3). \tag{2.42}$$

Last, but not least, a quantity is called a *tensor field* (a tensor field of rank two) if it has nine components that transform at the change of the reference system according to the law:

$$\bar{\xi}_{ij}(\bar{x}_1,\bar{x}_2,\bar{x}_3) = \frac{\partial x_k}{\partial\bar{x}_i}\frac{\partial x_s}{\partial\bar{x}_j}\,\xi_{ks}(x_1,x_2,x_3). \tag{2.43}$$

2.3.2   Stress tensor

Consider a body B, at a given instant t, and a closed surface S inside B. We would like to investigate the interaction of the matter inside S with that outside S.

Let us consider a small portion of the surface S, called ΔS; n is the unit normal of ΔS. Let us assume that the matter outside ΔS exerts on the matter inside ΔS a force $\Delta \mathbf{F}$. In non-polar materials, when ΔS tends to zero, the ratio $\Delta S/\Delta \mathbf{F}$ tends to a finite value, and the resulting moment of the forces acting on ΔS calculated in a generic point of ΔS tends to zero:

$$
\lim_{\Delta S \to 0} \frac{\Delta \vec{F}}{\Delta S} = \frac{\mathrm{d}\vec{F}}{\mathrm{d}S} = \overset{n}{\vec{T}}
$$

$$
\lim_{\Delta S \to 0} \frac{\Delta \vec{M}}{\Delta \vec{S}} = 0
$$

(2.44)

$\mathbf{T}$ is called the *stress vector*.

Every body that, for every surface S enclosed in it, fulfils Eq. (2.44), which is called the *Cauchy stress principle*, is called a *continuum*.

If we now consider the special case when the surface ΔS_k is orthogonal to the axis x_k of the Cartesian reference system (x_1, x_2, x_3), the stress vector associated with the surface ΔS_k has components in the direction of the reference system:

$$
\overset{k}{T_1} = t_{k1}; \quad \overset{k}{T_2} = t_{k2}; \quad \overset{k}{T_3} = t_{k3};
$$

(2.45)

The three components of each of the three stress vectors (one for each surface normal to an axis of the Cartesian reference system) can be represented by a 3×3 matrix. By convention, we usually note the components normal to the surface with sigma, and those parallel with tau:

$$
t_{ij} = \begin{array}{l} \sigma_i \to i = j \\ \tau_{ij} \to i \neq j \end{array}.
$$

(2.46)

2.3.3 Strain tensor

Consider a segment of infinitesimal length inside a body B, connecting the point $P(a_1, a_2, a_3)$ to its neighbor $P'(a_1 + \mathrm{d}a_1, a_2 + \mathrm{d}a_2, a_3 + \mathrm{d}a_3)$. The length of the segment is:

$$
\mathrm{d}s_0^2 = a_{ij}\,\mathrm{d}a_i\,\mathrm{d}a_j,
$$

(2.47)

where a_{ij} is the geometric function for the reference system a_i. When B is deformed, the infinitesimal segment inside it takes a length

$$
\mathrm{d}s^2 = g_{ij}\,\mathrm{d}x_i\,\mathrm{d}x_j,
$$

(2.48)

where g_{ij} is the function of Euclidean metrics for coordinate system x_i. Applying the transformation law we can write:

$$
\mathrm{d}s_0^2 = a_{ij}\frac{\partial a_i}{\partial x_l}\frac{\partial a_j}{\partial x_m}\,\mathrm{d}x_l\,\mathrm{d}x_m,
$$

(2.49)

$$\mathrm{d}s^2 = g_{ij} \frac{\partial x_i}{\partial a_l} \frac{\partial x_j}{\partial a_m} \mathrm{d}a_l \, \mathrm{d}a_m. \tag{2.50}$$

It is possible to show that the difference of the squared lengths is:

$$\mathrm{d}s^2 - \mathrm{d}s_0^2 = \left(g_{\alpha\beta} \frac{\partial x_\alpha}{\partial a_i} \frac{\partial x_\beta}{\partial a_j} - a_{ij} \right) \mathrm{d}a_i \, \mathrm{d}a_j \tag{2.51}$$

or in the other reference system:

$$\mathrm{d}s^2 - \mathrm{d}s_0^2 = \left(g_{ij} - a_{\alpha\beta} \frac{\partial a_\alpha}{\partial x_i} \frac{\partial a_\beta}{\partial x_j} \right) \mathrm{d}x_i \, \mathrm{d}x_j. \tag{2.52}$$

We define the strain tensors:

$$E_{ij} = \frac{1}{2} \left(g_{\alpha\beta} \frac{\partial x_\alpha}{\partial a_i} \frac{\partial x_\beta}{\partial a_j} - a_{ij} \right), \tag{2.53}$$

$$e_{ij} = \frac{1}{2} \left(g_{ij} - a_{\alpha\beta} \frac{\partial a_\alpha}{\partial x_i} \frac{\partial a_\beta}{\partial x_j} \right), \tag{2.54}$$

so that

$$\mathrm{d}s^2 - \mathrm{d}s_0^2 = 2E_{ij} \, \mathrm{d}a_i \, \mathrm{d}a_j, \tag{2.55}$$

$$\mathrm{d}s^2 - \mathrm{d}s_0^2 = 2e_{ij} \, \mathrm{d}x_i \, \mathrm{d}x_j, \tag{2.56}$$

The tensor E_{ij} is called the *Green strain tensor*, while the tensor e_{ij} is called the *Cauchy strain tensor* or *Almansi strain tensor*. The Green strain tensor expresses the deformation in the reference system of the deformed body, where the Almansi strain tensor expresses the deformation in the reference system of the undeformed body.

Now let us describe the deformation of the body in a single rectangular (rectilinear and orthogonal) Cartesian reference system; in this case, the function of the metrics will remain unchanged during the deformation. We define the *displacement vector* **u**:

$$u_i = x_i - a_i \rightarrow i = 1,2,3. \tag{2.57}$$

We can write

$$\begin{aligned} \frac{\partial x_\alpha}{\partial a_i} &= \frac{\partial u_\alpha}{\partial a_i} + \delta_{\alpha i} \\ \frac{\partial a_\alpha}{\partial x_i} &= \delta_{\alpha i} - \frac{\partial u_\alpha}{\partial x_i} \end{aligned} \tag{2.58}$$

Under these conditions it is possible to demonstrate that:

$$E_{ij} = \frac{1}{2}\left[\frac{\partial u_j}{\partial a_i} + \frac{\partial u_i}{\partial a_j} + \frac{\partial u_\alpha}{\partial a_i}\frac{\partial u_\alpha}{\partial a_j}\right]$$
$$e_{ij} = \frac{1}{2}\left[\frac{\partial u_j}{\partial x_i} + \frac{\partial u_i}{\partial x_j} - \frac{\partial u_\alpha}{\partial x_i}\frac{\partial u_\alpha}{\partial x_j}\right]$$
(2.59)

Now if we can assume *small deformations*, the derivatives of the displacement vector are small and the terms with the product of two derivatives can be neglected:

$$E_{ij} = \frac{1}{2}\left[\frac{\partial u_j}{\partial a_i} + \frac{\partial u_i}{\partial a_j}\right]$$
$$e_{ij} = \frac{1}{2}\left[\frac{\partial u_j}{\partial x_i} + \frac{\partial u_i}{\partial x_j}\right]$$
(2.60)

If in addition to small deformations we can assume *small displacements*, then the derivatives of the displacement with respect to the deformed and undeformed reference system would be negligibly different, which would mean that $E_{ij} = e_{ij}$. Under these assumptions, and using the notations $\mathbf{u} = (u, v, w)$ and $\mathbf{x} = (x, y, z)$, the *infinitesimal strain tensor* can be written as:

$$
\begin{aligned}
e_{xx} &= \frac{\partial u}{\partial x} & e_{xy} &= \frac{1}{2}\left(\frac{\partial u}{\partial y} + \frac{\partial v}{\partial x}\right) & e_{xz} &= \frac{1}{2}\left(\frac{\partial u}{\partial z} + \frac{\partial w}{\partial x}\right) \\
e_{yx} &= e_{xy} & e_{yy} &= \frac{\partial v}{\partial y} & e_{yz} &= \frac{1}{2}\left(\frac{\partial v}{\partial z} + \frac{\partial w}{\partial y}\right) \\
e_{zx} &= e_{xz} & e_{zy} &= e_{yz} & e_{zz} &= \frac{\partial w}{\partial z}
\end{aligned}
$$
(2.61)

2.3.4 Theory of elasticity: Hooke's law

The materials for which stress and strain tensors are linearly dependent are called *Hookian* materials. Hooke's law is valid for these materials, in its generalized tensor form, proposed by Cauchy:

$$\sigma_{ij} = D_{ijkl}e_{kl},$$
(2.62)

where σ_{ij} is the stress tensor, e_{kl} is the strain tensor (the Almansi strain tensor when deformations cannot be assumed small) and D_{ijkl} is called the *elasticity tensor*. Being symmetric in ij, and in kl, of the 81 elements of this tensor only 36 are independent. If we can postulate the existence of a *strain energy density* function:

$$W = \frac{1}{2}D_{ijkl}e_{ij}e_{kl},$$
(2.63)

so that

$$\frac{\partial W}{\partial e_{ij}} = \sigma_{ij},\qquad(2.64)$$

which is usually the case, then we can assume that the quadratic form defining the strain energy is also symmetric, which means that $D_{ijkl} = D_{klij}$, and this brings the number of independent constants down to 21. So in the worst case we can predict the elastic behavior of a material when we have measured experimentally 21 independent elastic constants. Fortunately, most materials exhibit additional symmetries, which significantly reduce this number. For example, if the material behavior is symmetric with respect to three orthogonal planes, the number of independent constants goes down to nine, and if the material is *isotropic* (has the same properties in all directions) there are only two constants to measure.

2.3.5   Navier equation

The most common problem we face in solid mechanics is the prediction of the stresses and strains in every point of the body B, knowing the forces and constraints acting on the body, its geometry, and the elastic constants of the material that form it.

If we can assume linear elasticity and material isotropy, we combine the equations of motion, the equations that define the deformation as a function of the displacements, the equation of continuity, which ensures that mass balance is preserved, Newton's second law, and, of course, Hooke's law into a single equation called the *Navier equation*:

$$G\frac{\partial^2 u_i}{\partial x_j^2} + (\lambda + G)\frac{\partial^2 u_i}{\partial x_j \partial x_i} + X_i = \rho\frac{\partial^2 u_i}{\partial t^2},\qquad(2.65)$$

where λ and G are the elastic constants, ρ is the density, and X_i is the vector of the volume forces (i.e. gravitational forces).

Unfortunately, this equation is extremely difficult to solve in any but the simplest cases, where the body and the boundary conditions have such a simple geometry that very strong implications can be made. For a generic body subjected to a generic system of forces, the only possibility is to formulate the problem numerically.

2.3.6   The numerical solution of the elasticity problem

The Navier equation does not allow an easy numerical solution. So before we can use numerical methods, we need to reformulate the elasticity problem in a more convenient way.

Let us consider a body B, of volume V, and external surface S. The body is subject to a system of external actions – conservative and in static equilibrium – which we can divide into surface actions $\mathbf{T}$, and volume actions $\mathbf{F}$. We want to determine the stresses and the strains in each point of the body. Since both stresses and strains can be expressed in terms of displacements, all we need is to determine how the system of action deforms the body to produce a displacement field $\mathbf{u}(x,\ y,\ z)$. From the displacement we can then calculate the deformation and the stress.

Since we are in a conservative system, the equilibrium configuration is characterized by a minimum of the total potential. Thus, to find the equilibrium displacement vector, we need to solve this differential equation:

$$\mathbf{u} \to \frac{\partial \Pi}{\partial \mathbf{u}} = 0, \tag{2.66}$$

where Π is the total potential of the body B, defined as:

$$\Pi = \int_V W \, \mathrm{d}V + \int_V G \, \mathrm{d}V + \int_S g \, \mathrm{d}S. \tag{2.67}$$

The first integral expresses the total deformation energy, the second the work of volume forces, and the third the work of the surface forces. The variation of this form (called a functional) is:

$$\partial \Pi = \int_V \partial W \, \mathrm{d}V + \int_V \partial G \, \mathrm{d}V + \int_S \partial g \, \mathrm{d}S. \tag{2.68}$$

The variation of a functional is called the *variational form*. The solution to our problem is the displacement vector field that minimizes Eq. (2.68). If all forces are conservatives we can express them through their potentials:

$$\begin{aligned} \mathbf{F} &= \frac{\partial G}{\partial \mathbf{u}} \\ \mathbf{T} &= \frac{\partial g}{\partial \mathbf{u}} \end{aligned}. \tag{2.69}$$

We already saw that for the strain energy density it is possible to write a similar equation:

$$\frac{\partial W}{\partial e_{ij}} = \sigma_{ij}, \tag{2.70}$$

from which we obtain

$$\partial \Pi = \int_V \sigma_{ij} \partial e_{ij} \, \mathrm{d}V - \int_V \mathbf{F} \partial \mathbf{u} \, \mathrm{d}V - \int_S \mathbf{T} \partial \mathbf{u} \, \mathrm{d}S = 0 \tag{2.71}$$

so the elasticity problem is now reduced to finding the displacement vector field that minimizes this variational form. This formulation more easily allows a numerical solution.

2.3.7   Galerkin method

Of the many possible methods of transforming the minimization of the variational into a numerically treatable problem, I present here the Galerkin method, which is also valid under conditions less restrictive than those imposed previously (conservative, static equilibrium).

Let us assume that our elastic problem can be expressed by a differential equation of form:

$$L\mathbf{u} = \mathbf{P}, \tag{2.72}$$

where L is a generic differential form, while $\mathbf{P}$ is the vectorial field that express the boundary conditions. Let us generate an approximation of the displacement vector field as a weighted piece-wise interpolation of some base functions:

$$\hat{\mathbf{u}} = \sum_{i=1}^{n} a_i f_i. \tag{2.73}$$

In general, since these interpolations provide only an approximation of the true displacement field, we can write:

$$L\hat{\mathbf{u}} - \mathbf{P} = e(x) \neq 0. \tag{2.74}$$

Now let us use the same base functions to define a weighting function:

$$\phi = \sum_{i=1}^{n} \phi_i f_i. \tag{2.75}$$

The approximated displacement field that satisfies the following equation gives the solution to the elastic problem:

$$\hat{\mathbf{u}} \rightarrow \min[\textstyle\int_V \phi (L\hat{\mathbf{u}} - \mathbf{P})\mathrm{d}V] = \min[e(x)]. \tag{2.76}$$

The solution minimizes $e(x, y, z)$, properly weighted, over the entire body volume.

2.3.8   The finite-element method

The most common method of solving solid mechanics problems is the finite-element method. This method was first developed empirically by engineers in need of solving complex elasticity problems, and only afterward formalized in the frame of the Galerkin variational methods. There are many ways to explain the method; the one that I have found particularly effective in teaching computational biomechanics to biomedical engineers retains the basic virtual work approach.

As a first step we need to transform all key quantities into vector form. Consider a body B, subjected to a system of volume forces, a system of surface forces, and a *system of concentrated forces*:

$$\mathbf{f}^B = \begin{bmatrix} f_x^B \\ f_y^B \\ f_z^B \end{bmatrix}; \qquad \mathbf{f}^S = \begin{bmatrix} f_x^S \\ f_y^S \\ f_z^S \end{bmatrix}; \qquad \mathbf{R}_C^i = \begin{bmatrix} R_{cx}^i \\ R_{cy}^i \\ R_{cz}^i \end{bmatrix}. \tag{2.77}$$

The displacement with respect to the undeformed configuration in each point of the body is:

$$\mathbf{u}(x,y,z) = \begin{bmatrix} u \\ v \\ w \end{bmatrix}. \tag{2.78}$$

Let us assume that both the deformations and the displacements are small. Considering its symmetries, for which only six of the nine elements are independent, the infinitesimal strain tensor can be written in vector form as:

$$\boldsymbol{\varepsilon}^{\mathrm{T}} = [\varepsilon_x \; \varepsilon_y \; \varepsilon_z \; \gamma_{xy} \; \gamma_{yz} \; \gamma_{zx}], \tag{2.79}$$

where

$$\varepsilon_x = \frac{\partial u}{\partial x}; \qquad \varepsilon_y = \frac{\partial v}{\partial y}; \qquad \varepsilon_z = \frac{\partial w}{\partial z}$$

$$\gamma_{xy} = \frac{\partial u}{\partial y} + \frac{\partial v}{\partial x}; \qquad \gamma_{yz} = \frac{\partial v}{\partial z} + \frac{\partial w}{\partial y}; \qquad \gamma_{zx} = \frac{\partial w}{\partial x} + \frac{\partial u}{\partial z}. \tag{2.80}$$

Similarly, we can write the stress tensor as a six-component vector:

$$\boldsymbol{\sigma}^{\mathrm{T}} = [\sigma_x \; \sigma_y \; \sigma_z \; \tau_{xy} \; \tau_{yz} \; \tau_{zy}], \tag{2.81}$$

Hooke's law can be written as:

$$\boldsymbol{\sigma}^{\mathrm{T}} = \underline{\underline{E}} \cdot \boldsymbol{\varepsilon}, \tag{2.82}$$

where now E is a 6×6 matrix called the *elasticity matrix*. For an isotropic material, the nine non-null elements can all be expressed as a function of the Young modulus E and the Poisson ratio v:

$$\begin{aligned} E_{11} &= E_{22} = E_{33} = (1-v)c \\ E_{12} &= E_{23} = E_{13} = vc \\ E_{44} &= E_{55} = E_{66} = G \end{aligned}, \tag{2.83}$$

where

$$\begin{aligned} c &= \frac{E}{(1+v)(1-2v)} \\ G &= \frac{E}{2(1+v)} \end{aligned}. \tag{2.84}$$

If we assume that the body is in equilibrium, the principle of virtual work tells us that for every virtual displacement field that is compatible with the boundary conditions and the continuum assumption, the work of the internal forces must be equal to that of the external forces:

$$\int_V \mathbf{u}^\mathrm{T} \mathbf{f}^B \, \mathrm{d}V + \int_S \mathbf{u}^{S^\mathrm{T}} \mathbf{f}^S \, \mathrm{d}S + \sum_i \mathbf{u}^{i^\mathrm{T}} \mathbf{R}_C^i = \int_V \varepsilon^\mathrm{T} \sigma \mathrm{d}V. \qquad (2.85)$$

The three terms on the left side of the equation are the virtual work of the volumetric, surface, and concentrated forces, respectively. Their sum must be equal to the total deformation energy that is stored inside the body. Now let us elaborate this equation with respect to the displacements, as the most common formulation of the method requires.

Let us assume that the body is completely divided into many non-overlapping sub-volumes, which we shall call *finite elements*. Each finite element is a cell identified by a given number of vertices that we call *nodes*. For the generic element m we define a local reference system (ξ, ψ, ζ). We can provide an approximation of the displacement field for the portion of the body limited by the volume of the element m:

$$\hat{\mathbf{u}}^{(m)}(\xi, \psi, \zeta) = \underline{\underline{H}}^{(m)}(\xi, \psi, \zeta) \cdot \mathbf{U}, \qquad (2.86)$$

where H is the displacement interpolation matrix and $\mathbf{U}^\mathrm{T} = [U_1, V_1, W_1, U_2, V_2, W_2, \dots U_n, V_n, W_n]$ is the vector of the displacements of the n elements' nodes. Definition of the displacement field as an interpolation of the nodal displacements is possible only if the finite elements have certain shapes, and are properly connected together. We shall consider element of type h, for which the order of the interpolation polynomials is low and constant, and for which the convergence toward each displacement field is obtained by progressively reducing the size of the elements. On the contrary, elements of type p converge to the exact solution by keeping the element size fixed, but increasing the order of the interpolating polynomials.

Similarly, we can express the approximation to the deformation field in the region of the body inside the element m as an interpolation of the nodal displacements:

$$\hat{\varepsilon}^{(m)}(\xi, \psi, \zeta) = \underline{\underline{B}}^{(m)}(\xi, \psi, \zeta) \cdot \mathbf{U}, \qquad (2.87)$$

where B is the displacement–deformation matrix. Finally, we can write the equation of elasticity for the approximations of stress and strain fields as:

$$\hat{\sigma}^{(m)} = \underline{\underline{E}}^{(m)}(\xi, \psi, \zeta) \cdot \hat{\varepsilon}^{(m)}. \qquad (2.88)$$

So if we know the various matrices and the nodal displacements we can compute approximations of displacement, stress, and strain fields in every point inside element m. To have the same information over B, we need to assemble all the local solutions we obtain for each element into a global solution that is valid for the entire body. For a body decomposed in finite elements, the virtual work equation can be written as:

$$\sum_m \int_{V^{(m)}} \tilde{\mathbf{u}}^{(m)^{\mathrm{T}}} \mathbf{f}^{B(m)} \, \mathrm{d} V^{(m)} + \sum_m \int_{S_1^{(m)}, S_2^{(m)}, \ldots, S_q^{(m)}} \tilde{\mathbf{u}}^{S(m)^{\mathrm{T}}} \mathbf{f}^{S(m)} \, \mathrm{d} S^{(m)} + \sum_i \tilde{\mathbf{u}}^{i^{\mathrm{T}}} \mathbf{R}_C^i ,$$
$$= \sum_m \int_{V(m)} \tilde{\boldsymbol{\varepsilon}}^{(m)^{\mathrm{T}}} \boldsymbol{\sigma}^{(m)} \mathrm{d} V^{(m)} \tag{2.89}$$

where each integral is limited to the region associated to element m, and the total work is the summation over all elements. Using these interpolation equations, we can express stress, strain, and displacements as a function of the sole nodal displacements:

$$\tilde{\mathbf{U}}^{\mathrm{T}} \left[\sum_m \int_{V^{(m)}} \underline{\underline{\mathbf{B}}}^{(m)^{\mathrm{T}}} \underline{\underline{\mathbf{E}}}^{(m)} \underline{\underline{\mathbf{B}}}^{(m)} \, \mathrm{d} V^{(m)} \right] \mathbf{U}$$
$$= \tilde{\mathbf{U}}^{\mathrm{T}} \left[\sum_m \int_{V^{(m)}} \underline{\underline{\mathbf{H}}}^{(m)^{\mathrm{T}}} \mathbf{f}^{B(m)} \, \mathrm{d} V^{(m)} + \sum_m \int_{S_1^{(m)}, S_2^{(m)}, \ldots, S_q^{(m)}} \underline{\underline{\mathbf{H}}}^{S(m)^{\mathrm{T}}} \mathbf{f}^{S(m)} \, \mathrm{d} S^{(m)} + \sum_i \mathbf{R}_C^i \right],$$
$$\tag{2.90}$$

or more synthetically:

$$\tilde{\mathbf{U}}^{\mathrm{T}} \underline{\underline{K}} \mathbf{U} = \tilde{\mathbf{U}}^{\mathrm{T}} (\mathbf{R}_B + \mathbf{R}_S + \mathbf{R}_C). \tag{2.91}$$

The matrix K is called the *stiffness global matrix*. The product of the interpolation matrices is integrated over each element volume, to produce a stiffness local matrix. All these local matrices are then assembled in the global matrix K. The terms on the right side, within parentheses, are the contribution of the volume, surface, and concentrated force systems, to the global forces' resultant $\mathbf{R}$. Since the equation must be true for every virtual nodal displacement, it can be true if and only if

$$\underline{\underline{K}} \mathbf{U} = \mathbf{R}. \tag{2.92}$$

This is called the fundamental equation of the finite-element method. It tells us that the nodal displacement vector and the resultant of all force systems are related through the stiffness matrix. By solving this system of algebraic equations we can determine an approximation of the displacement field at the nodes, and from that use the interpolation functions to compute an approximation of the displacements, stresses, and strains in every point of the body. The degree of approximation is related to the size of the elements we use to decompose the body, for two reasons. Firstly, the stiffness matrix approximates the true body stiffness as the number of elements tends to infinity (and thus the element size tends to zero); secondly, the smaller the element, the more likely it is that the finite-order polynomials will be able to interpolate accurately the spatial gradients of displacement, stress, and strain over the element volume.

Of course the choice of polynomials is important. Let us consider an element m whose displacement interpolation functions are:

$$\mathbf{u}^{(m)} = \underline{\underline{N}}_s^{(m)} \cdot \mathbf{U}. \tag{2.93}$$

Of course, we can also express the coordinates of each point inside the element as an interpolation of the nodal coordinates:

$$\mathbf{x}^{(m)} = \underline{\underline{N}}_c^{(m)} \cdot \mathbf{X}, \tag{2.94}$$

assuming in both cases that the same interpolation matrix yields a number of advantages. These element formulations are called *isoparametric*:

$$\underline{\underline{N}}_c^{(m)} = \underline{\underline{N}}_s^{(m)} = \underline{\underline{N}}^{(m)}. \tag{2.95}$$

The interpolation matrix N is expressed with respect to the local coordinate system (ξ, η, ζ) called *natural*, so that the element nodes always have null or unitary coordinates. Similarly the faces that delimit the finite element are characterized by the values $\xi = \pm 1$, $\eta = \pm 1$, and $\zeta = \pm 1$.

For example, in an isoparametric hexahedral eight-node element the coordination interpolation functions take the form

$$N_i = \frac{1}{8}(1 + \xi\xi_i)(1 + \eta\eta_i)(1 + \zeta\zeta_i), \tag{2.96}$$

where ξ_i, η_i, ζ_i are the natural coordinates of node i. The interpolation is thus linear in each coordinate. With this element formulation, the displacements are

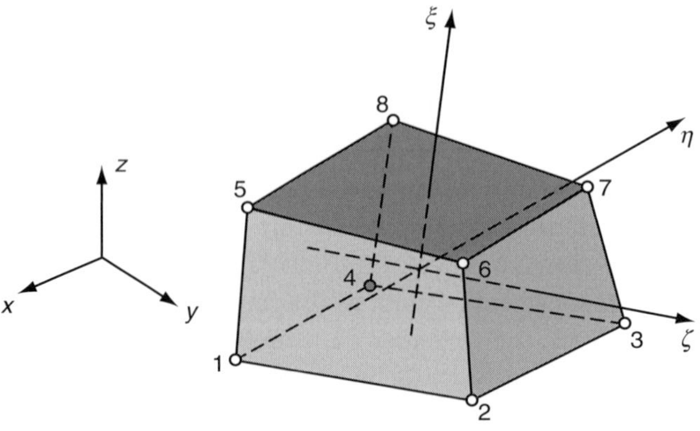

Figure 2.4 Eight-node isoparametric finite element and its natural reference system.

interpolated linearly over the element, and the deformation components are also linearly interpolated over the displacements.

The determination of the local stiffness matrix involves the solution of this integral:

$$\underline{\underline{K}}^{(m)} = \int_{V^{(m)}} \underline{\underline{B}}^{(m)^{\mathrm{T}}} \underline{\underline{E}}^{(m)} \underline{\underline{B}}^{(m)} \, \mathrm{d}V^{(m)}. \tag{2.97}$$

Assuming that the elasticity matrix is known, the only problem is the displacement–deformation matrix, which can be derived from the definition of deformation, rewriting it in matrix form:

$$\varepsilon^{(m)} = \underline{\underline{d}}_e \mathbf{u} = \underline{\underline{d}}_e \underline{\underline{N}}^{(m)} \cdot \mathbf{U}, \tag{2.98}$$

where the matrix differential operation is defined as

$$\underline{\underline{d}}_e^{\mathrm{T}} = \begin{bmatrix} \dfrac{\partial}{\partial x} & 0 & 0 & \dfrac{\partial}{\partial y} & 0 & \dfrac{\partial}{\partial z} \\[2ex] 0 & \dfrac{\partial}{\partial y} & 0 & \dfrac{\partial}{\partial x} & \dfrac{\partial}{\partial z} & 0 \\[2ex] 0 & 0 & \dfrac{\partial}{\partial z} & 0 & \dfrac{\partial}{\partial y} & \dfrac{\partial}{\partial x} \end{bmatrix}. \tag{2.99}$$

Remembering that

$$\varepsilon^{(m)} = \underline{\underline{B}}^{(m)} \cdot \mathbf{U}, \tag{2.100}$$

it follows that

$$\underline{\underline{B}}^{(m)} = \underline{\underline{d}}_e \underline{\underline{N}}^{(m)}. \tag{2.101}$$

Since the interpolation matrices in the isoparametric elements are expressed with respect to the local reference system, the matrix **B** is also expressed locally. Thus, the stiffness matrix must be integrated with respect to this local reference system. But we can transform the reference system inside an integral by using the *Jacobian matrix*:

$$\underline{\underline{J}} = \begin{bmatrix} \dfrac{\partial \xi}{\partial x} & \dfrac{\partial \xi}{\partial y} & \dfrac{\partial \xi}{\partial z} \\[2ex] \dfrac{\partial \eta}{\partial x} & \dfrac{\partial \eta}{\partial y} & \dfrac{\partial \eta}{\partial z} \\[2ex] \dfrac{\partial \zeta}{\partial x} & \dfrac{\partial \zeta}{\partial y} & \dfrac{\partial \zeta}{\partial z} \end{bmatrix} \tag{2.102}$$

with which the stiffness matrix equation becomes:

$$\underline{\underline{K}}^{(m)}(x,y,z)=\iiint \underline{\underline{B}}^{(m)^{\mathrm{T}}}\,\underline{\underline{E}}^{(m)}\,\underline{\underline{B}}^{(m)}\,\mathrm{d}x\,\mathrm{d}y\,\mathrm{d}z$$
$$=\iiint \underline{\underline{B}}^{(m)^{\mathrm{T}}}\,\underline{\underline{E}}^{(m)}\,\underline{\underline{B}}^{(m)}\,\underline{J}\,\mathrm{d}\xi\,\mathrm{d}\eta\,\mathrm{d}\zeta \tag{2.103}$$

Except for the simplest elements, this integral must be solved numerically, using *Gaussian quadrature*. This method returns an exact value only in the sampling points, called *Gauss points*. This is why the most accurate estimates of the stresses and strains inside the element are provided at corresponding Gauss points.

2.3.9 Non-linear problems

Everything described so far relies on the assumption of linearity. However, in many practical problems the assumptions that yield this linearity produce errors too large to be acceptable. In these cases we should go back to the most fundamental theories, and try to work out a solution under much more complex conditions. While this is the only option in some cases, there is an alternative. If we can assume that the elastic problem is linear locally (in space or time), we can still use the results discussed previously, in the frame of an iterative scheme.

In a system like:

$$\underline{\underline{K}}\mathbf{U}=\mathbf{R}, \tag{2.104}$$

the system is linear if both **K** and **R** are independent of time and of **U**. In non-linear systems, we can relax this restriction, and accept that either the stiffness matrix or the boundary conditions are functions of time or of **U**. In this case, the numerical solution can be found with an iterative scheme like the *Newton–Raphson scheme*.

If a system is in equilibrium we can write

$$\sum_{m}\int_{V^{(m)}}\tilde{\mathbf{u}}^{(m)^{\mathrm{T}}}\mathbf{f}^{B(m)}\,\mathrm{d}V^{(m)}+\sum_{m}\int_{S_1^{(m)},S_2^{(m)},\dots,S_q^{(m)}}\tilde{\mathbf{u}}^{S(m)^{\mathrm{T}}}\mathbf{f}^{S(m)}\,\mathrm{d}S^{(m)}+\sum_{i}u^{i^{\mathrm{T}}}R_C^{i}$$
$$=\sum_{m}\int_{V^{(m)}}\tilde{\boldsymbol{\varepsilon}}^{(m)^{\mathrm{T}}}\boldsymbol{\sigma}^{(m)}\mathrm{d}V^{(m)} \tag{2.105}$$

The terms on the left side express the virtual work of the external forces **R**; in the right term we replace the virtual deformation with its interpolation over the virtual nodal displacements. We obtain

$$\tilde{\mathbf{U}}^{\mathrm{T}}\mathbf{R}=\tilde{\mathbf{U}}^{\mathrm{T}}\left[\sum_{m}\int_{V^{(m)}}\underline{\underline{B}}^{(m)^{\mathrm{T}}}\boldsymbol{\sigma}^{(m)}\,\mathrm{d}V^{(m)}\right]. \tag{2.106}$$

The term inside the square brackets defines the vector of the *internal* forces **F**, i.e. the nodal forces that we must apply to each element in order to equilibrate the stresses inside the element. Under equilibrium conditions we can thus write:

$$\mathbf{R} - \mathbf{F} = 0. \tag{2.107}$$

For the sake of illustration, let us assume that our system is non-linear owing to its dependency on time. If the phenomenon is time-varying, we can always assume a *dynamic equilibrium*, where the system is in each instant t in a (different) equilibrium condition:

$$^t\mathbf{R} - {}^t\mathbf{F} = 0. \tag{2.108}$$

Let us assume that the solution at time t is known; we want to compute the solution at time $t + \Delta t$. As a first approximation we can pretend that the stiffness matrix and the nodal displacement at time $t + \Delta t$ remain unchanged from time t:

$$^{t=\Delta t}\mathbf{K}^{(0)}\Delta\mathbf{U}(1) = {}^{t+\Delta t}\mathbf{R} - {}^{t+\Delta t}\mathbf{F}^{(0)}. \tag{2.109}$$

Here we use the right superscript to indicate the iteration step of the solution. Assuming that $\mathbf{K}$ and $\mathbf{F}$ did not change from the previous time step we calculate the changes in the nodal displacement vector that would equilibrate the external force vector. We then use this first estimate of the variation of the nodal displacement vector to compute a second estimate of the nodal displacement vector at time $t + \Delta t$:

$$^{t+\Delta t}\mathbf{U}^{(1)} = {}^{t+\Delta t}\mathbf{U}^{(0)} + \Delta\mathbf{U}^{(1)}. \tag{2.110}$$

With this new estimate of the nodal displacement vector, we compute a new estimate of the internal forces vector and of the stiffness matrix:

$$^{t+\Delta t}\mathbf{F}^{(1)} = \sum_m \int_{V^{(m)}} \underline{B}^{(m)^{\mathrm{T}}} E^{(m)} B^{(m)\,t+\Delta t}\mathbf{U}^{(1)}\,\mathrm{d}V^{(m)}, \tag{2.111}$$

$$^{t+\Delta t}\mathbf{K}^{(1)} = \frac{\partial^{t+\Delta t}\mathbf{F}^{(1)}}{\partial^{t+\Delta t}\mathbf{U}^{(1)}}. \tag{2.112}$$

Now we can generalize the Newton–Raphson iteration scheme for step (i):

$$^{t+\Delta t}\mathbf{K}^{(i-1)}\Delta\mathbf{U}^{(i)} = {}^{t+\Delta t}\mathbf{R} - {}^{t+\Delta t}\mathbf{F}^{(i-1)}, \tag{2.113}$$

$$^{t+\Delta t}\mathbf{U}^{(i)} = {}^{t+\Delta t}\mathbf{U}^{(i-1)} + \Delta\mathbf{U}^{(i)}. \tag{2.114}$$

The iteration continues until the difference between the external forces $\mathbf{R}$ and the internal forces $\mathbf{F}$ or the variation of the displacement vector $\Delta\mathbf{U}$ becomes smaller than the convergence tolerance.

2.3.10 Numerical solution methods

The solution of the elastic problem can thus be reduced to the solution, within an iterative cycle, of a system of linear equations of type

$$\mathbf{K} \cdot \mathbf{u} = \mathbf{F}. \tag{2.115}$$

The trivial solution of this problem is to invert the stiffness matrix:

$$\mathbf{u} = \mathbf{K}^{-1} \cdot \mathbf{F}. \qquad (2.116)$$

Unfortunately, inverting a matrix costs $O(N^3)$ operations where N is the number of unconstrained degrees of freedoms. As modern finite-element models can easily exceed 300 000 degrees of freedom, the inversion would take 10^{16} operations. With a 1 Gflops processor (one billion operations per second) this would require in the most ideal conditions 10^7 seconds, or 115 days of calculations to solve a single model.

Thus, it is clear that we need to find more efficient ways to solve the problem. Finite-element solvers belong to three large families: *sparse, frontal,* and *iterative.*

In most cases, the stiffness matrix is sparse, i.e. it has many null elements. Usually these null elements are randomly distributed in the matrix, but in many cases it is possible to reorder the nodes and the elements so that all non-null elements are located in a band, mostly at a distance B from the matrix diagonal; B is called the bandwidth of the matrix. For very sparse matrices, and for efficient renumbering algorithms, the bandwidth can become much smaller than N. The matrix is then divided into sub-matrices, and the Gauss elimination is applied computing only the elements inside the band. A sparse solver requires $O(NB^2)$ operations. In favorable conditions a stiffness matrix can have a sparseness of 90%, which can be reduced to a bandwidth of only 50 elements. This would bring down to 75 seconds the solution time of the 300 000 degrees of freedom model on the 1 Gflops processor.

There are certain cases where the matrix bandwidth cannot be sensibly reduced by numbering. In these cases sparse solvers are not efficient, and frontal solvers are preferred. In these solvers the assembly of the stiffness matrix and the Gauss elimination are performed simultaneously. The active calculus matrix is composed, adding equations element by element, and the relative known terms. When all local matrices that affect a degree of freedom have been added, Gauss elimination is performed for that degree of freedom.

Another critical factor in solving very large systems of linear equation is the storage of the stiffness matrix. If the computer has enough fast access memory (RAM) to store the active portion of the matrix (in-core solution), this is not a problem. But if the matrix is so large that part of it has to be stored in a much slower disk memory (swap) (out-of-core solution) then the performance impact of reading and writing data from a memory that can be a thousand times slower can be significant. In these cases it is frequently more convenient to use an iterative solver. These solvers initialize the problem with a tentative solution and then iteratively improve it locally, which enables only portions of the stiffness matrix to be loaded into the memory, with the portion size dynamically determined on the basis of the available fast access memory, optimizing the solution to the available hardware. A series of procedures that undergoes the collective name of *pre-conditioning* can drastically improve the efficiency and reduce the memory footprint of iterative solvers, especially over very large problems, by pre-processing the matrix elements.

2.3.11 Further reading

Continuum mechanics and the finite-element method are subjects of a very large body of literature, whose exhaustive review is beyond the scope of this book. For an extensive discussion on solid mechanics I recommend the classic Fung and Tong (2001). Some of the most classic references on the finite-element method are Fung and Tong (2001); Strang and Fix (2008); Bathe and Wilson (1976); Zienkiewicz and Taylor (2005); and Cook (1995).

2.4 The meshless-cell method

2.4.1 Basic formulation

In many biomedical applications the source of information on the shape of the body, organ, or tissue is a 3D image produced by computed tomography, magnetic resonance imaging, or other biomedical imaging method. To transform a 3D image into a finite-element model it is generally necessary to segment the structure of interest in the image, extract the mathematical closed surface that best fits the outer boundary of the segmented volume, and then decompose the volume enclosed by this surface into a solid finite-element mesh. These operations, usually referred as *pre-processing*, tend to be laborious, and require a significant amount of time by a skilled operator to be performed. Thus, it is no surprise that computational biomechanics researchers are constantly looking for methodological alternatives that considerably reduce or skip entirely the duration of the pre-processing phase. As an example, I present the *meshless-cell method* (MCM), which our group has been developing in collaboration with the University of Trieste.

The cell method (CM) is a numerical method based on the direct discrete formulation of field equations. An exhaustive description can be found in Tonti (2001). While traditional numerical methods are based on differential formulations of physical laws, in the cell method we directly write the algebraic equations referring to finite regions of the studied domain.

The cell method uses global variables referring to the geometrical elements of two cell complexes: a *primal cell complex*, defined by creating a mesh of the studied geometry; and a *dual cell complex*, obtained by arbitrarily dividing (e.g. using the Voronoi criterion) the primal cells in parts ascribed to its nodes. Dual cells are tributary regions of nodes of the primal cell complex. The solution is obtained for the nodes of the primal cell complex by writing a balance equation on the dual cell of each node. As an example, for each node h, the balance equation of the elastostatic problem takes the form:

$$\sum_{c \in \mathfrak{I}(h)} \mathbf{T}_h^c + \mathbf{F}_h = 0, \tag{2.117}$$

where $\Im(h)$ is the set of primal cells sharing the generic node h; $\mathbf{F}_h$ is the volume force acting on the dual cell of node h; and $\mathbf{T}_h^c$ is the force acting on the face of the dual cell of node h that belongs to the primal cell c. This is possible in the mesh-based approach because the boundary of the dual cell can be divided into faces, each falling in one and only one cell of the primal complex.

The total surface force acting on the boundary of the dual cell is the sum of the forces acting on each face. These forces can be expressed in terms of displacements of the primal cell nodes: polynomial interpolation of the primal cell's node values makes it possible to express the cell strain tensor in terms of nodal displacements: $\underline{\varepsilon}^c = \mathbf{B}^c \mathbf{u}^c$. The constitutive law of the material of each primal cell produces the relationship between the strain and stress tensors: $\underline{\sigma}^c = \mathbf{D}^c \underline{\varepsilon}^c$; finally, the Cauchy relation expresses the force $\mathbf{T}_S^c$ acting on the surface S in terms of the stress tensor components: $\mathbf{T}_S^c = A_S \underline{\sigma}^c$. All these relations can be combined to express the surface force $\mathbf{T}_S^c$ in terms of displacements of primal nodes of the cell containing the surface:

$$\mathbf{T}_S^c = A_S \mathbf{D}^{\,c} \mathbf{B}^{\,c} \mathbf{u}^c. \tag{2.118}$$

Then, Eq. (2.117) can be transformed into:

$$\sum_{c \in \Im(h)} \mathbf{K}_h^c \mathbf{u}^c + \mathbf{F}_h = 0, \tag{2.119}$$

which shows that the displacements of the primal cell nodes can be obtained by solving the global system derived from the assembling balance, Eq. (2.119), for all nodes h.

2.4.2   Meshless formulation

In its classical implementation, the cell method still requires the domain to be decomposed into a mesh. However, it is possible to imagine a meshless approach where this requirement is relaxed, or to be more accurate where only local meshes, generated automatically, are required.

The core of the cell method is the balance equation referred to tributary regions of each primal node. The validity of a balance equation is general, regardless of the shape or extension of the tributary region; tributary regions can overlap but, globally, they have to cover the whole domain. It is possible to define these tributary regions even in the absence of a mesh comprising the whole domain (Zovatto and Nicolini, 2003; 2006) but relying only on a local mesh that can be built automatically by connecting each primal node with some of its surrounding nodes, chosen according to a certain criterion. The tributary region of each node is then the dual cell defined on its local mesh. For a generic node h (conventionally named the "pole") the procedure can be summarized in the following steps:

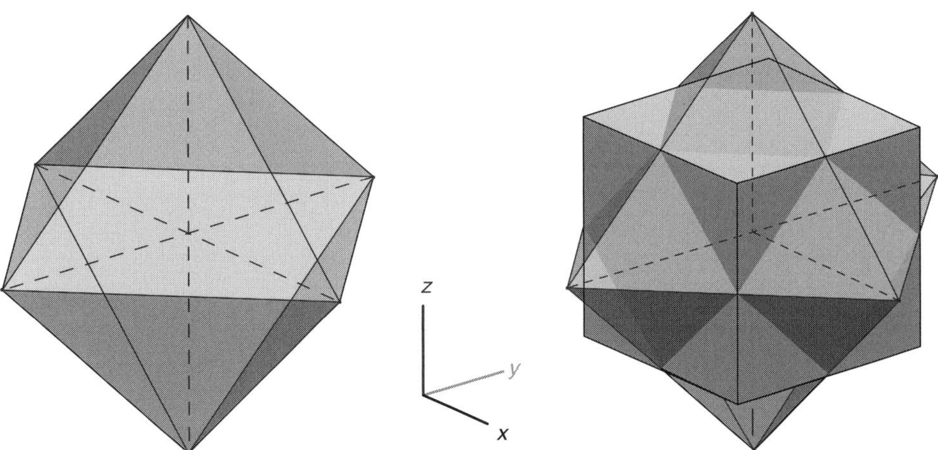

Figure 2.5 On the left, the local mesh consisting of eight tetrahedra for a generic internal node, built to interpolate the displacement field. On the right, the tributary region of the same node is also shown in gray, corresponding to the CT voxel. With permission from Elsevier (Taddei *et al.*, 2008).

- Select N neighbor nodes, named "satellites;" the number of satellites is arbitrary but they have to enable the building of a local mesh surrounding the pole;
- Build a local mesh surrounding the pole node, connecting the pole with the satellites;
- Define the tributary region of the pole;
- Write the balance equation referred to this tributary region;
- Assemble, node by node, a local balance equation in the global system.

In our implementation, the local mesh is generated by defining a set of linear tetrahedra between each pole node and the nearest nodes in the biomedical 3D image lattice, as in the left side of Figure 2.5.

The tributary region of each node is the voxel corresponding to the node itself, as shown in the right side of Figure 2.5. For each tributary region the balance equation, Eq. (2.117), must be written in terms of primal node displacements. However, unlike the CM mesh-based approach, the boundary of each tributary region cannot be divided into faces that are pertinent to a uniquely defined cell of the primal complex, since a local mesh is automatically built at run-time for each node.

To express the surface forces, i.e. the stresses acting on the surface, in terms of nodal displacement, all possible local elements incident on the same face of the tributary region boundary should be identified. This may not be a trivial problem, and we did not attempt to solve it generally. However, in the present MCM implementation, which relies on a structured grid organization of the primal node cloud, this could easily be obtained, since each face of the boundary of the tributary region could be associated with only two different tetrahedra (Figure 2.6).

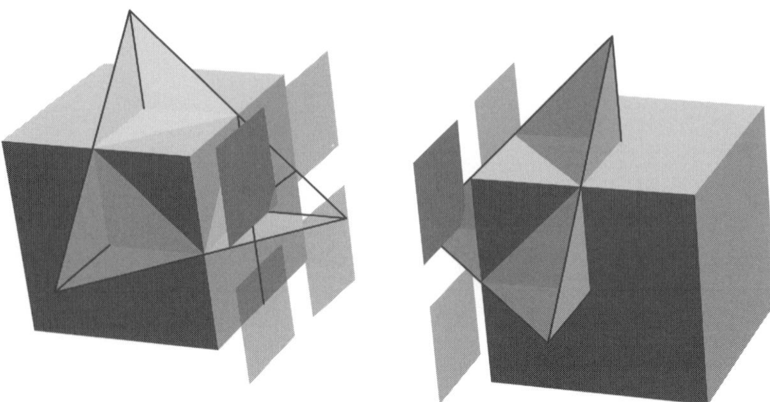

Figure 2.6   Two adjacent tributary regions artificially separated to show the piece of boundary surface (dark square) associated with two different local mesh tetrahedra of the primal complex. The force acting on this surface is calculated from the average between the two stress fields predicted by each single tetrahedron. With permission from Elsevier (Taddei *et al.*, 2008).

Thus, the surface force acting on the shared face is calculated by simply averaging the forces predicted by the two tetrahedra. The global system of linear equations, generated by assembling the local balance equations, is then solved numerically.

At native resolution a CT dataset is formed by hundreds of 512×512 two-dimensional images. Assuming a 512-slice dataset, we would have 134 million voxels; if we generate a MCM node for each voxel, with three translational degrees of freedom, this would produce over 400 million unconstrained degrees of freedoms. This is true for most of these alternative methods: we avoid part or all the pre-processing operator time, at the cost of a significantly increased solution time. This latter problem can be minimized using parallel solvers and multiprocessor or cluster architectures that considerably reduce the wall-clock time to solve such models.

The method described here is currently under evaluation both for organ-level and tissue-level applications. A first validation study (Taddei *et al.*, 2008) showed that MCM predicts the strains measured on the surface of a proximal human femur with an average error of 16%, whereas the conventional finite-element models produce an average error of only 9% for the same problem. However, in this study hardware limitations prevented the simulation from being run at the full-image resolution. More recent results on models run at full-image resolution confirm that the MCM method is as accurate as the finite-element method, but without the massive pre-processing effort that the latter requires.

3   The body level

A description of the anatomy and the physiology of the neuromusculoskeletal system, and the methods used to model the musculoskeletal dynamics, in particular to predict the muscle and joint forces acting on the skeleton during a given movement.

3.1   Introduction

The musculoskeletal apparatus is an organ system whose main functions are the support of the body, the provision of motion, and the protection of vital organs. During these functions bones are subjected to considerable internal forces, transmitted primarily by the muscles and the joints, and to external forces transmitted through the skin and the other connective tissues. Thus, to determine the forces acting on the skeleton we need to investigate how the whole neuromuscular apparatus works during physiological and para-physiological activity. This can be done by modeling the whole body during these movements, with methods that help us to estimate which forces are transmitted to the skeleton instant by instant. The scope of this chapter is to describe these whole-body modeling methods.

3.2   Elements of anatomy and physiology

3.2.1   Descriptive anatomy

This book is intended for practitioners of musculoskeletal biomechanics; nevertheless, in this and in the following sections I shall review some functional anatomy, physiology, and biology, to ensure a common background and a common terminology.

The musculoskeletal system provides form, stability, and movement to the human body. It is tightly interrelated to the nervous, vascular, and integumentary (skin) systems. Its primary components are bones, cartilage, muscles, ligaments, and tendons.

In all vertebrates, the skeleton is made of *bones*, organs with a mineralized extracellular matrix; but this is only one of the strategies chosen by evolution to ensure shape and support. For example platyhelminthes, nematodes, and annellids have a hydrostatic skeleton, made of vesicles filled with pressurized fluid. Bones are much stiffer than any other tissue, and provide a framework of sufficient rigidity. The average adult human skeleton typically has 206 bones (Gray, 1918).

The bones are connected by *joints*. Joints are divided into *synarthroses*, in which the osseous components are united by fibrous tissue or cartilage, and *diarthroses*, in which the opposing bone ends are separated by a cavity filled with fluid. Synarthroses are practically immovable, whereas in most cases diarthroses are movable joints, with a variable degree of mobility depending on the anatomic location of the joint. Diarthrotic joints are complex structures where two or more bone extremities (epiphyses), wrapped by hyaline cartilages and connected by ligaments, articulate one against the other inside capsules of ligaments and fasciae, filled with synovial fluid. Joints oppose minimal resistance when the bones they couple move in given directions, and much higher resistance when these bones move in other directions. Bones and joints form the skeleton, which can be seen as an articulated frame.

The skeleton can provide protection and eventually support (although in reality any static posture other than lying supine also requires some muscle activity to be maintained) but the generation of movement also involves the *muscles*. These should be more properly defined as *myotendinous units* or neuromyotendinous units, as skeletal muscles are generally composed of a motor unit (what we usually call the muscle), connected to the bones via tendons, and activated through the neuromuscular junction. The functional unit of a skeletal muscle is the motor unit. It consists of a motor neuron with its cell body in the ventral horn of the spinal cord and its peripheral axon, the neuromuscular junction, and the muscle fibers innervated by the neuron. The ventral motor neuron is the final common pathway conducting neural impulses from the central nervous system to the muscle. When an action potential arrives at the axon terminal, voltage-dependent calcium channels open and calcium ions flow from the extracellular fluid into the motor neuron's intracellular fluid. This influx of calcium ions triggers a biochemical cascade that causes the synaptic vesicles, filled with neurotransmitters, to fuse to the motor neuron's cell membrane and release acetylcholine into the synaptic cleft. Acetylcholine diffuses across the synaptic cleft and binds to the nicotinic acetylcholine receptors that dot the motor end plate. These receptors are ligand-gated ion channels that open when bound by acetylcholine, allowing sodium ions to flow into the muscle's intracellular fluid. This ionic flux produces a local depolarization of the muscle fibers facing the synaptic cleft (the motor end plate); this is known as the end-plate potential. The depolarization propagates on the fiber surface, and then penetrates it via the *transverse tubules* (T-tubules), where it opens voltage-dependent calcium channels that allow the calcium ions contained in the *sarcoplasmic reticulum* (an organelle of the muscle cell) to diffuse into the intracellular space and activate the muscle contraction.

Tendons connect the muscle motor unit to the bone, and the contraction of muscles is transformed into movement of the skeleton. Tendons are bands of fibrous connective tissue, formed by parallel arrays of collagen fibers, mostly made of type 1 collagen. The tendon's microstructure is crimped, and this produces a non-linear force–displacement characteristic; at lower forces the tendon is less stiff and deforms more easily, while at higher forces it becomes stiffer. The tendon's stiffness is also a function of strain rate; the tissue is less stiff at slower deformations, and becomes stiffer as the strain rate increases. Tendons are strong: in physiological conditions the most common cause of failure is tendon avulsion, where it is not the tendon that fails, but rather the bony insertion.

3.2.2 Elements of muscle physiology

In the body there are three types of muscle: *smooth muscle*, which covers the intestinal walls and the blood vessels, and which exhibits a totally involuntary contractile behavior; *cardiac muscle*, which is similar to smooth muscle but exhibits time-deterministic contractile behavior; and *skeletal muscle*, which is the muscle of interest here. Skeletal muscles have a hierarchical structure: a layer of connective tissue called the *epimysium* wraps the muscle, which is composed of bundles called *fascicles*, each of which is wrapped in another layer of connective tissue, called the *perimysium*. The fascicle is composed of elongated cells called *muscle cells* or *muscle fibers*; a third layer of connective tissue, called the *endomysium*, wraps each fiber, which is composed of cylindrical organelles called *myofibrils*.

Muscle fibers are filiform, multinucleate cells with diameters ranging between 0.01 and 0.1 mm, and a length that can exceed 300 mm. They are delimited by a cellular membrane, called the *sarcolemma*, and filled with intracellular fluid, called *sarcoplasm*, within which are immersed the cell nuclei, usually located at the periphery of the cell, and the myofibrils. Each muscle fiber is innervated, can be stimulated by a nervous impulse, can propagate its potential, and consequently contracts. Some muscle fibers, usually whitish, contract more quickly, while others, usually reddish, contract more slowly. Each muscle includes both fast and slow fibers, although in different proportions.

Myofibrils are composed of repeating contractile units known as *sarcomeres*. Each sarcomere consists of thick and thin myofilaments, whose arrangement is largely responsible for the cross-striated banding pattern observed under light and electron microscopy. Sarcomeres are delineated at each end by a region called the *Z line* where thin filaments of *actin* extending from opposite directions meet and link together. Inside the sarcomere, the actin filaments are intercalated with thicker *myosin* filaments; it is the interaction between these two molecules that contract the fibril when calcium ions are released into the sarcoplasm. The myofibrils are anchored to the sarcolemma by the cytoskeletal network; the sarcolemma is connected to the endomysium, which is connected to the perimysium, which is connected to the epimysium, which in turn is connected to the tendon and through it to the bone. Thus, when the myofibrils contract, the force is transmitted from the contractile cell to the bone insertion.

When an activation signal propagates through a motor neuron, the muscle fibers it innervates contract and then relax. If the neuron propagates repeated bursts of activation, the fibers will repeatedly contract and relax; however, if the activation signal is transmitted with the frequency above a certain threshold, the muscle fibers do not have enough time to relax before the next contraction. This threshold frequency is called *tetanic*, and the force the muscle fibers express at that frequency (tetanic force) is the highest possible.

From a functional point of view, force, length, and contractile velocity are strongly correlated; in general in experiments we keep one of these quantities constant, control a second, and measure the third. We are experimenting under *isometric* conditions when the muscle length is kept constant, *isotonic* conditions when the force is kept constant, and *isokinetic* condition when the velocity is kept constant (*isostatic* when the velocity is zero). We define the contraction of a shortening muscle as *concentric* and the contraction of a lengthening muscle as *eccentric*. It should be noted that these terms are used in biomechanics with slightly different meanings.

3.2.3   Activation dynamics

A muscle consists of thousands of muscle fibers organized into *motor units*, which comprise a group of muscle fibers, often several hundred, which are innervated by a single motor neuron. A motor unit is activated in an all-or-none fashion by a single action potential, which travels from the motor neuron along the axon to the muscle fibers.

A single action potential produces a brief contraction of the muscle fiber called a *twitch*, whose duration depends on the muscle fiber type. The durations of both the contraction and relaxation phases of the twitch are longer for slow-twitch (type I) than fast-twitch (type II) fibers. In skeletal muscle the contraction times (time to peak) range from 7.5 ms for fast extra-ocular muscle fibers, 40 ms for intermediate, to 90 ms for slow muscle fibers (Figure 3.1).

All fibers in a given motor unit are of the same type, being determined by the nature of the motor neuron. Small tonically active motor neurons prompt development of slow-twitch types; large, phasic motor neurons favor fast-twitch fibers.

The force that a motor unit can express is a function of the frequency of activation (firing rate) of the innervating motor neuron. The firing rate is defined as the number of action potentials per second. The force produced by each muscle fiber, innervated by the motor neuron, increases with firing rate because of the accumulation of intracellular calcium (Ca^{2+}), whose concentration produced by a single

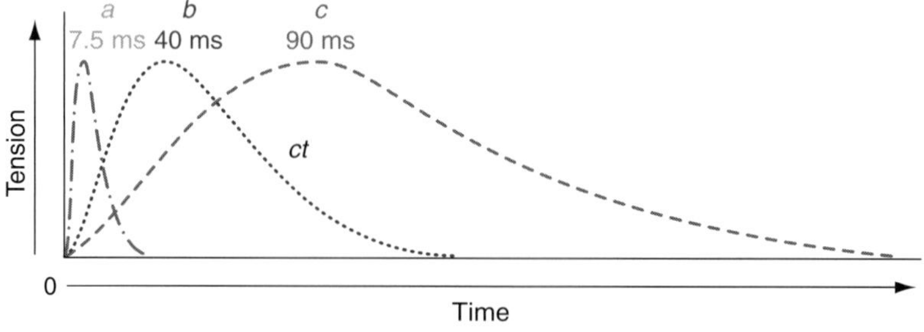

Figure 3.1   Contraction times (time to peak tension) for (a) extra-ocular, (b) fast-twitch skeletal, and (c) slow-twitch skeletal muscle fibers.

action potential, increases and decreases more rapidly than the isometric twitch force. Therefore, the amount of force added by a second action potential occurring immediately after the first will depend on the time interval between them, e.g. on the amount of intracellular calcium at the time of occurrence. The additional force contribution by a second action potential drops steeply as a function of the interval between two successive action potentials.

If a motor unit is activated at a steady frequency, the force will initially rise and then oscillate around a mean value at the frequency of activation, producing what is called an *unfused tetanus*. Both the mean force and the initial rate of force development will increase as the firing rate increases. The higher the firing rate, the smaller the oscillation with respect to the mean force will be. At high firing rates, there is no noticeable oscillation in force. This smooth steady force is called *tetanus*. Type I motor units reach tetanus at lower frequencies because they have longer twitch contraction times than type II.

Human motor units can be voluntarily activated at instantaneous firing rates of about 100 Hz during short forceful contractions. The maximum firing rates that they can sustain during steady contractions are considerably lower and generally do not exceed 30 Hz.

When a muscle is activated voluntarily under isometric conditions, motor units tend to become active in a fixed order. The recruitment order is correlated with the amount of force that a motor unit can produce. The motor unit force is related to the number of muscle fibers and the size of the muscle fibers that it comprises. The motor unit that produces the smallest force is recruited first. It remains active and the next motor unit is recruited as the total muscle force increases. The motor units that produce the largest forces are the last to be recruited. As the total muscle force increases, each newly recruited unit contributes an increment in force, which is a similar percentage of the total muscle force. In this way the force can be increased smoothly (Figure 3.2).

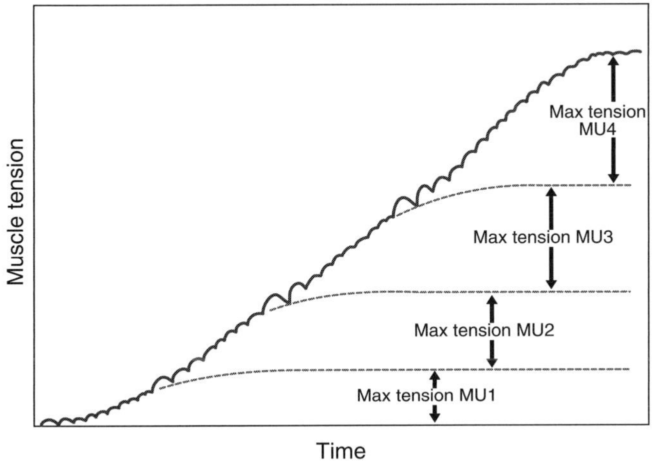

Figure 3.2 Recruitment of motor units (MU) is progressive, so as to allow a smooth build-up of muscle tension.

The different mixtures of fiber type composing each muscle, however, are not sufficient to explain the large range of possible mechanical characteristics of skeletal muscles; functionally, an important feature of muscles is how the fibers are organized in space. There are muscles with parallel fibers, fusiform muscles, and muscles where the fibers are not parallel to the line of action; the latter are called *pennate* muscles. Some muscles have a single pennation direction, whereas others have two or more. Different muscle architectures can provide very different mechanical properties that allow very precise and reactive muscle units (e.g., hand muscles) or slow and very strong muscle units (e.g., hip muscles). A functional determinant of the muscle ability to produce forces is the maximum diameter at rest; in most of the literature the *physiological cross-sectional area* (PCSA) is preferred as an indicator of muscle size. If the number and quality of contractile fibers per unit area is constant, the PCSA is proportional to the tetanic force that a muscle can express. However, it should be noted that the number and the quality of the fibers in a muscle can vary considerably; with the appropriate training program we can increase a muscle force by 100%, with negligible increases in the PCSA (Jones *et al.*, 2008). The maximum isometric stress is reported to vary between 0.35 and 1.37 MPa (Buchanan *et al.*, 2004).

3.2.4   Motor control

Movement is possible because the central nervous system controls the activation of the muscles in a coordinated way; this is called *motor control*. A detailed discussion of human motor control is beyond the scope of this book. Here I mention only a few basic concepts, which we shall need in the rest of this chapter.

A motion is usually initiated by a sensorial input, a sensation. From this perception to the actual activation of the muscles typically takes around 200 milliseconds. This relatively long time is used to execute:

- *Sensation*, including neural transmission from sensory receptors in the eyes, ears, etc., to the brain, which takes about 15 ms.
- *Perception*, including the retrieval of long-term memories to organize, classify, and interpret the sensations, which takes about 45 ms.
- *Response selection*, which takes about 75 ms.
- *Response execution* of an action plan (a step-by-step sequence of events that make up the planned movement), which takes about 15 ms.

As a first approximation we can describe two distinct types of motor control: closed-loop and open-loop. Closed-loop motor control uses perception to adjust muscle movements consciously and continuously. Each stimulus–response adjustment involves the entire sequence of events listed, and thus takes at least 200 milliseconds. When this time is excessive, the body uses another type of motor control, in which the sequence of muscle activations is executed according to a pre-planned trajectory in the activation pattern space. In reality, the motor control mechanisms are much more complex. We can perform a high-level goal with very good repeatability, but if we analyze how muscles are activated we shall see that each repetition involves a significant degree of variability. In other words, motor control appears to be a stochastic process, but only within the redundant control space that is irrelevant to the high-level goal (Todorov and Jordan, 2002).

One point that is relevant for this book is that there is strong evidence that one of the low-level goals of motor control is the protection of the skeleton. The so-called Charcot's joint provides indirect evidence. This is a well-defined radiological condition that results from a fairly large number of diseases, including syphilis and diabetes, all inducing neurological damage resulting in partial or complete loss of perception at the affected joint (Pakarinen *et al.*, 2002). As a result, the joint under-goes fairly rapid and massive destruction, with little or no pain (again because of the lack of perception). There are two theories on this disease (Brower and Allman, 1981): for the first, the lack of sensation removes a vital proprioceptive signal that helps us in protecting our joints during daily motor activities; without this protec-tion, normal physiological movement is sufficient to destroy the joint. The second theory postulates that the neurological damage also produces an inflammatory condition; the excess of blood reaching the joint accelerates the physiological cel-lular processes associated with bone overloading, producing significant bone loss. However, it is important to notice that both theories recognize that the loss of local perception produces dramatic overloading. This confirms our theory that motor control aims to protect bones and joints from damage during daily activities. Thus, to investigate skeletal overloading we should understand in which cases such pro-tective control mechanisms fail in full or in part. This is only possible using whole-body modeling.

3.3 Whole-body modeling

3.3.1 Modeling assumptions

As explained in Chapter 1, models must always include some degree of idealization of the observed phenomenon. In whole-body modeling the skeleton is idealized as a kinematic chain, muscles are idealized as bundles of linear actuators connected only to two bones, and the rest of the body is considered only in terms of its mass, which is lumped in selected points. I shall first revise these assumptions, and then I shall develop the modeling approach using them.

A kinematic chain is the assembly of one or more kinematic pairs (joints) linking two or more rigid bodies. These joints in general allow unresisted motion in certain directions (called degrees of freedom), and totally prevent motion in others. Thus, idealizing the skeleton with a kinematic chain involves assuming that:

- Bones are infinitely rigid,
- Joints present no resistance to free directions of motion,
- Joints are infinitely rigid in the blocked directions of motion.

Simple calculations – using beam theory and knowing that during normal physiological activities long bone deformation produces strains of the order of 2×10^{-3} – suggest that the assumption of rigid bones produces errors by neglecting changes in the frame configuration during motion of the order of a few millimeters. Thus, we can consider this assumption acceptable in most cases. Of course, like any other idealization, this one also remains acceptable only within a range of conditions. In the case of high-force activities, deformations might become much larger; if the specific model is sufficiently sensitive to configuration as to produce large differences in predictions for displacement errors of the order of millimeters; if the total motion of the bone segments is comparable to that due to deformation (as in the spine (Shirazi-Adl, 1994)); in all these cases the errors produced by this idealization might become unacceptable. It is worth noting that this particular idealization is the fruit of the reductionist approach to modeling. Bone deformation happens at the organ scale, whereas motion is usually described at the body scale. Classic modeling methods can deal effectively with only one scale at a time; thus, to remove the interdependency with the organ scale, we must postulate that bone shape will not change during motion. In Chapter 7 we shall see how body–organ multiscale modeling methods can help to remove this particular idealization.

The second idealization introduces only negligible errors. In vertebrates there is no advantage in wasting energy by producing frictional heat each time a joint articulates; thus it is not surprising that the functional anatomy of vertebrate joints is a masterpiece of low-friction articulation. The coefficient of kinetic friction for a human joint is around 0.01; while that for ice-on-ice friction is 0.03 (Serway, 1995).

On the other hand, the third idealization has frequently been questioned. The problem is that human joints have to ensure stability but also flexibility. Inside joints, the resistance to certain degrees of freedom is produced firstly by soft connective tissues, such as cartilage, ligaments, or menisci, and only for much larger ranges by stiffer bony surfaces. Thus in general, during motion human joints show movement along ideally constrained directions, but usually of amplitude much smaller than that measured along unconstrained directions. Whether neglecting these residual movements over resisted degrees of freedom is acceptable or not depends on the specific scope of the model, its construction, the required precision, etc.

More complex is the assumption that muscles can be idealized as bundles of linear actuators connected to only two bones. Muscles are three-dimensional masses that significantly change shape during contraction. This contraction is highly anisotropic, and it is strongly affected by the orientation of the muscle fibers within the muscle volume (pennation). In long, fusiform muscles, during contraction the resultant muscle force vector remains more or less parallel to the conjunction of the centers of the muscle origin and insertion areas; but in short, multipennated muscles this is not generally true (Rohrle and Pullan, 2007). This problem may be mitigated by using a clever representation of the muscle volume by multiple actuators, each representing a primary pennation (contraction) direction. Another problem with this idealization is that it neglects the transversal interaction of the muscle with the other connective structures surrounding it. For example, Peter Hujing and colleagues (Yucesoy *et al.*, 2003) have convincingly documented the intermuscular myofascial force transmission. As in most cases, these modeling idealizations may or may not be acceptable, depending on the scope of the modeling activity. A major problem we face in this context is, however, the lack of a directly measurable standard against which models' predictions can be validated. The available experimental methods let us record the motion of the body (Cappozzo *et al.*, 2005), measure the forces between the body and the external world during motion (e.g., ground reaction forces during perambulation), or record electromyography (EMG) signals related to the muscle action potentials, providing insights on how the various muscles are activated during the motion (Kleissen *et al.*, 1998). If a patient has a joint prosthesis, we can embed into the prosthetic device force sensors that transmit their real-time recordings, enabling the joint forces transmitted to bones during various activities to be measured (Bergmann *et al.*, 2001; Damm *et al.*, 2010). However, we have no method of measuring directly and non-invasively the force exerted by each muscle during a certain movement – the information that is needed to validate the musculoskeletal model directly, and also the best way to determine whether a given idealization within the model produces acceptable errors.

Another problem is that of neuromotor variability. Even the quantities we can measure (ground force, EMG, etc.) present considerable variability when the same subject repeats the same motor task in apparently the same way. Which of these values should be used to validate our model? In the 1960s Bernstein concluded from a large body of experimental observations of functional motor tasks that at higher levels of the nervous system the spatial aspects of the requisite movements are controlled rather than the action of specific joints or muscles (Bernstein, 1967). In the 1980s Pietro Morasso found that in point-to-point reaching tasks, repeated tests showed good regularity in the hand movement trajectories, but a great variability of joint trajectories (Morasso, 1981). These and other observations supported the development of the "uncontrolled manifold" (UCM) theory (Scholz and Schoner, 1999), which suggests that motor control strategies focus on the goal of the task, and that every trajectory within the manifold of the task-equivalent configuration of the muscle actuators is virtually possible. This theory is worth mentioning here because it suggests that the variability we observe when the same subject repeats the same task is somehow built into the motor control mechanism.

Thus, it is extremely difficult to decide when the idealizations described here produce acceptable errors and when they do not. This forces researchers to formulate the problem in totally different terms, which will be exposed in the following sections. But before this, we need to describe how these musculoskeletal models are built.

3.3.2   Modeling musculoskeletal dynamics

The central nervous system decides to perform a voluntary movement; neurons activate muscles; muscles produce forces; these forces accelerate joints; so that the body performs the desired movement. If the skeletal system is represented by a kinematic chain with n degrees of freedom q, actuated by m muscles, the relationship between the movement and muscle forces can be written as:

$$\mathbf{M}(q)\ddot{q} + \mathbf{C}(q)\dot{q} + \mathbf{G}(q) + \mathbf{R}(q) \times \mathbf{F}^{M} + \mathbf{M}_{e}(q,\dot{q}) = 0, \qquad (3.1)$$

where: q, $\dot{q}$, and $\ddot{q}$ are the generalized coordinate vectors for position, velocity, and acceleration; $\mathbf{M}(q)$ is the system mass matrix ($n \times n$); $\mathbf{C}(q)$ is the matrix of the forces and moments produced by centrifugal and Coriolis actions; $\mathbf{G}(q)$ is the vector of the gravitational forces and moments; and $\mathbf{M}_{e}(q, \dot{q})$ is the matrix of the external forces and moments. The quantity $\mathbf{R}(q) \times \mathbf{F}^{M}$ represents the muscular joint moments, where $\mathbf{R}(q)$ is the matrix of the muscular moment arms ($n \times m$) and $\mathbf{F}^{M}$ are the myotendinous forces ($m \times 1$).

Equation (3.1) is usually redundant, in the sense that there are usually many more actuators than degrees of freedom ($m > n$). This means there are an infinite number of muscle force patterns that produce the same movement; the choice among them is taken by neuromotor control. One possibility we have is to presume that the neuromotor control works according to some optimization principle, and thus we can search among the infinite admissible solutions for that which minimizes a certain cost function.

To compute the muscle forces there are two alternative approaches: *forward dynamics* and *inverse dynamics* (Figure 3.3). In forward dynamics, $\mathbf{F}^{M}$ is assumed known and Eq. (3.1) is solved for q, $\dot{q}$, and $\ddot{q}$; in inverse dynamics, q, $\dot{q}$, and $\ddot{q}$ are assumed known, and Eq. (3.1) is solved for $\mathbf{F}^{M}$.

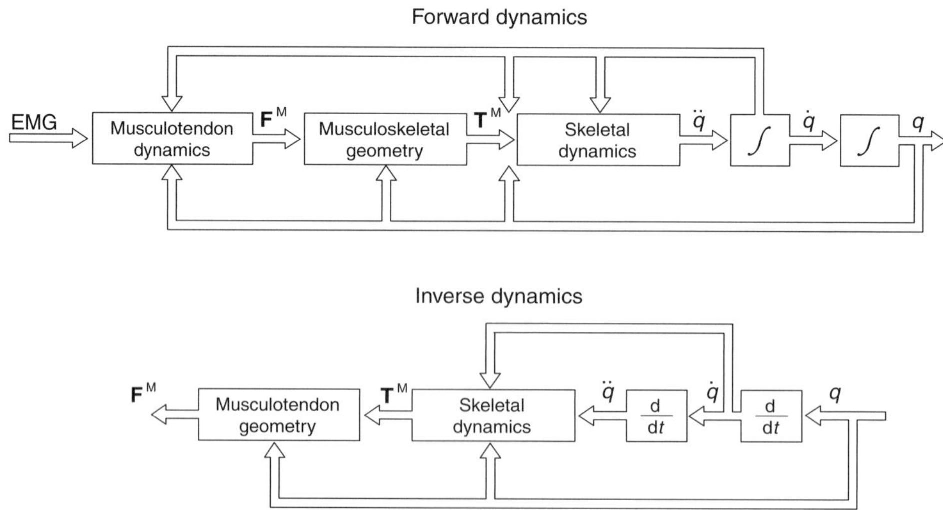

Figure 3.3   Schematic representations of the solution of the muscle forces problem with forward and inverse dynamics approaches. EMG = electromyography; $\mathbf{F}^{M}$ = muscle force; $\mathbf{T}^{M}$ = muscle torque; q, $\dot{q}$, $\ddot{q}$ = position, velocity, and acceleration of generalized coordinates.

In the forward dynamics approach, this optimization is usually done globally over the entire interval of time we want to model; on the contrary, with inverse dynamics it is possible to solve Eq. (3.1) for a single instant, using a local optimization scheme. But this is not mandatory: formulations of static optimization coupled with forward dynamics have also been provided to minimize tracking error of movements (e.g. (Thelen *et al.*, 2003)).

Assuming that we have a fully defined musculoskeletal geometry and a fully identified model of the musculotendon dynamics, the choice of approach seems to depend mostly on the type of data available as inputs. If the muscle excitation pattern during the movement is known (by EMG) or assumed, we can use the musculotendon dynamics model to predict each muscle force during the movement, and then use the musculoskeletal model to predict the position, velocity, and acceleration of each bone during motion. On the contrary, if the movement of the bones is known (by motion capture) or assumed, the musculoskeletal model can be used to predict the muscle forces and excitations during the movement. In practice, neither the bone movement nor the muscle activation patterns can be measured reliably. With motion capture we measure the motion of skin-attached markers, which are not rigidly fixed to the underlying bones. This produces so-called skin motion artifacts, which considerably reduce the accuracy with which we can measure the kinematics of bones during a movement. Direct non-invasive measurement of muscle forces is not possible; what we can do is to measure with electromyography the muscles activation patterns, and then use a musculotendon dynamics model to predict muscle forces from these activation patterns. But the experimental measurement of the muscles activation patterns is also somehow troubled. Needle electrodes produce discomfort in the subject being measured, whereas surface electrodes hardly allow recording the activation potential of deep muscles (very deep muscles are hard to reach with needle electrodes). The EMG measurements are usually very noisy, and the type of noise that affects them is difficult to separate from the signal (Clancy *et al.*, 2002). The activation potential evoked in one muscle can produce an artifact in the measurements made in another muscle nearby, owing to electric crosstalk phenomena (Farina *et al.*, 2004).

In the end the choice of forward or inverse dynamics is dictated by a number of factors, including the specific research question the model has to answer, the acceptable solution time, the availability and reliability of the experimental inputs, etc. When the goal of modeling is to predict the joint and muscle forces transmitted to the skeleton during movement, and if patient-specific motion data are available, inverse dynamics is usually the method of choice. I shall concentrate on this approach in the following pages.

3.3.2.1 Inverse dynamics modeling

The prediction of muscle forces using inverse dynamics involves a number of steps. Firstly, we need to define the patient-specific musculoskeletal model, put in the patient-specific movement data, and solve the skeletal dynamics by computing the vector of generalized forces that must be transmitted at each joint to achieve instantaneous equilibrium. If we know the inertial and dumping properties of the subject body, the motion kinematics, and the external moments and forces, we can compute the moments that the muscles must express at each joint to ensure instantaneous equilibrium:

$$\mathbf{R}(q) \times \mathbf{F}^{\mathrm{M}} = \mathbf{M}(q)\ddot{q} + \mathbf{C}(q)\dot{q} + \mathbf{G}(q) + \mathbf{M}_{\mathrm{e}}(q,\dot{q}). \tag{3.2}$$

Unfortunately, owing to the large number of muscles, there are infinitely many possible muscle activation patterns that ensure such instantaneous equilibrium. This is sometimes called the *neuromuscular indeterminacy problem*. However, if we assume that the neuromotor control works according to some optimal principle, we can find the unique solution to the constrained optimization problem, which selects from the infinite number of muscle force solutions the one that minimizes a certain cost function. The constraints emerge from the fact that, for each instantaneous configuration, each muscle bundle we modeled is capable of producing at most a tetanic force that depends on a number of muscle-specific parameters, as well as the actual length and the actual contractile velocity. Thus, in practice we also need a musculotendon dynamics model, even if it is not explicitly depicted in Figure 3.3 for the inverse dynamics approach.

3.3.2.2 Musculotendon dynamics model

Before we deal with the neuromuscular indeterminacy problem, we need to define the second model required to predict muscle forces: the musculotendon dynamics model. Such a model should define the relationship between the neuromotor activation signal, which we can normalize to range between zero and one, and the total force produced at the tendon by the muscle. As already mentioned, the biomechanics of a muscle can be modeled using a continuum model or a lumped-parameters model. Even if we limit our attention to this second group, a great variety of models are described in the literature; which should we use in our case? A key factor in this case is the information available; a number of models are very sophisticated, and require that we know a large number of parameters of each muscle anatomy and physiology. Some of these parameters can be determined only with highly invasive methods, and thus are limited to studies on isolated muscles, typically dissected post mortem. When the goal is to develop models usable in the clinical practice, we need to select those models that can be identified with data we can collect non-invasively, even at the cost of lower accuracy. As usual in these cases, a compromise might be to use only some subject-specific parameters, and supplement them with average values determined with large cadaveric studies. The relationships that link the neuromotor activation signal can be obtained by coupling the excitation dynamic model and the contraction dynamic model. In the following, both models will be discussed separately.

3.3.2.3 Single-fiber excitation dynamics

Muscle cannot be activated or relaxed instantaneously. The delay between muscle excitation and activation (or the development of muscle force) is due mainly to the time taken for calcium pumped out of the sarcoplasmic reticulum to travel down the T-tubule system and bind. The delay between muscle excitation (u, which represents the net neural drive) and muscle activation (a) can be written in the form of a first-order process:

$$\dot{a} = \frac{u - a}{\tau_a(a,u)},\tag{3.3}$$

where $\tau_a(a, u)$ is a time constant that varies with activation level and the excitation signal (Thelen *et al.*, 2003). Implicit in this equation is the assumption that muscle activation depends only on a single variable u. Other models separate the time constant during the rising and falling phases of activation.

3.3.2.4 Single-fiber contraction dynamics

Here we shall concentrate our attention on the lumped-parameters muscle dynamics model derived from the work of Archibald V. Hill, and then modified and refined over the years by other authors. During his studies on the thermodynamics of muscular contraction Hill proposed the following equation:

$$(v + b)(F + a) = b(F_0 + a),\tag{3.4}$$

where F represents the muscle force, v is the velocity of contraction, and a, b, F_0 are constants. If we assume that a and b are zero, Hill's equation suggests that muscle contraction occurs at constant power (the product of force and velocity); this means that the speed with which the muscle produces mechanical work through the conversion of chemical energy is constant. This assumption is reasonable in tetanic condition, which is indeed the condition produced in Hill's experiment. A frog's sartorius muscle is held at constant length L_0, and electrically stimulated with a growing frequency; the force will increase with the frequency, up to a value F_0, which is indeed the isometric tetanic force, which depends on the isometric length L_0. If we now suddenly release one of the two tendons of the muscle until the force is decreased to an imposed value F, the muscle shortens with a contraction velocity v that can be measured. Hill found that this velocity decreased hyperbolically as the force F was increased, in very good agreement with Eq. (3.4).

If we now define as v_0 the contraction velocity associated with the tetanic load F_0, Hill's equation can be normalized. If we set:

$$\begin{aligned} v_0 &= \frac{bF_0}{a} \\ c &= \frac{F_0}{a} \end{aligned},\tag{3.5}$$

Hill's equation can be written in dimensionless form as

$$\frac{v}{v_0} = \frac{1 - \dfrac{F}{F_0}}{1 + c\dfrac{F}{F_0}}.\tag{3.6}$$

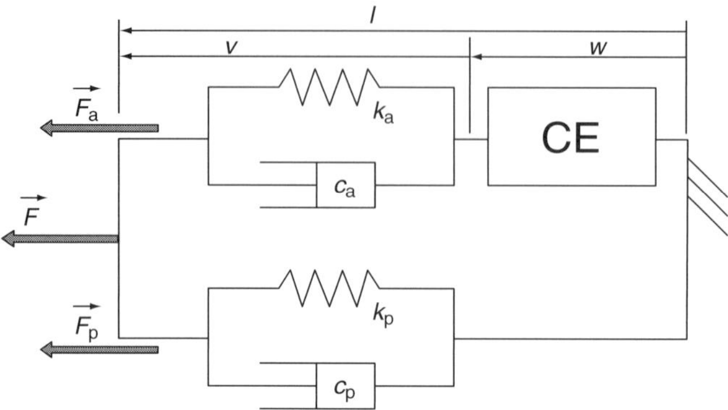

Figure 3.4   Hill's model of a muscle fiber. The total muscle force F is the sum of the active F_a and passive F_p components. The active component is a function of elongation v (via the stiffness k_a and the damping c_a), and of the contractile element CE, which elongates w. The passive component is a function of the total elongation u, via the stiffness k_p and the dumping c_p.

In this form, the constants identifying the equation are the tetanic force F_0 and the tetanic contraction velocity v_0, plus the constant c. All three are functions of the initial length L_0. The tetanic force becomes zero for very high and very low values of muscle lengthening; in the range where the tetanic force is not zero, it varies with the muscle lengthening in a characteristic way.

While Hill's original model described a very specific condition, over the years the model has been extended so that it can be used in a broader range of cases. In many cases these generalizations are referred to the single sarcomere, rather than the whole muscle; it is common to see the dynamic behavior of the muscle as a simple superposition of the dynamic behavior of its sarcomeres, which are assumed to be identical. The generalized Hill's model includes three elements (Figure 3.4): a viscoelastic element, which describes the passive behavior of the muscle, placed in parallel with two elements placed in series to each other. The first of these elements is the contractile element, with zero stiffness at rest; the second is the viscoelastic element that describes the behavior of the contracted muscle. However, the structural interpretation of these elements is not unique, and it is more appropriate to consider this model as purely phenomenological. A fourth element that describes the elastic behavior of the tendons is added when the model is used to describe a whole musculotendon unit.

The solution of the model is easy to derive if we consider at first the two elements of the active portion as a single one: we now have two elements in parallel, one active and the other passive. They both experience the same elongation u, but different forces F_a and F_p. If we now consider only the active branch, we have two elements in series, one contractile and the other viscoelastic, subject to the same force F_a but different displacements, v and w, so that $v + w = l$. Thus, for the two elements in parallel we can write:

$$F = F_p + F_a = (k_p l + c_p \dot{l}) + (k_a v + c_a \dot{v}), \tag{3.7}$$

where both the elastic and the viscous components are made evident for each force. For the two elements in series, considering that $v = u + w$, we can write:

$$F_a = k_a(l - w) + c_a(\dot{l} - \dot{w}) = F_c(\dot{w}, L_0). \tag{3.8}$$

The force expressed by the contractile element depends on the length of the muscle at the instant when the contraction is initiated L_0, and the velocity of the contractile element, or *contractile velocity* $\dot{w}$. L_0 is also called the *rest length;* since the contractile force is maximal at rest length; this is also called *optimal fiber length.*

To determine the force a muscle expresses in a given instant, and knowing the kinematic conditions of displacement and velocity, we need to determine the elastic and viscous constants of both active and passive elements, and the function that expresses the contractile force as a function of initial length and velocity. Currently, we have no established methods to measure these parameters non-invasively in vivo (although promising results are being obtained using magnetic resonance elastography (Bensamoun *et al.*, 2007) and ultrasound elastography (Hoyt *et al.*, 2008)). Thus, we are forced to resort to average values taken from the literature.

Under isotonic conditions (constant force) we know from Hill's experiments that the elongation velocity is maximum under zero force, that it rapidly decreases with the increase of the constant force applied, becoming zero when the force reaches the tetanic value (Figure 3.5b). Another set of experimental observations made on isolated sarcomeres (Horowitz and Pollack, 1993) showed how the isometric force strongly depends on the imposed length (Figure 3.5a). If we normalize the elongation to the rest length, the maximum isometric tetanic force remains constant in a range of imposed normalized elongation between 0.95 and 1.05, and then linearly decreases for increasing length, reaching the null value for elongations around 150%–170% of the rest length. If we shorten the muscle, the tetanic force is

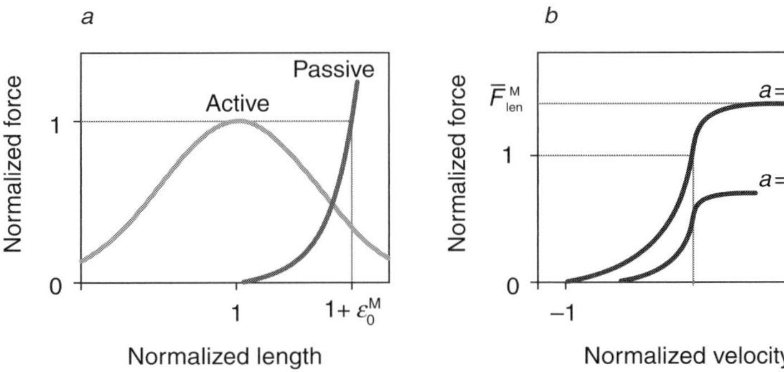

Figure 3.5 (a) Force–elongation curve for single muscle fiber. The active maximum tetanic force exerted by the fiber increases as the length increases up to the optimal fiber length, and then decreases; the passive force increases with the elongation up to the maximum fiber elongation ε_0^M. (b) Force–velocity curve for a single fiber. Under tetanic activation ($a = 1$) the muscle force increases with the velocity, up to a value F_{len}^M that is larger than the tetanic force in isometric conditions. Partial activation produces the same curve shape, but with smaller forces.

reduced more slowly for elongation down to 80% and then more rapidly to become null around an elongation of 50%–60% of the rest length. Conventionally, we can assume that a muscle can produce a contractile force in isometric conditions for elongations between 50% and 150% of the rest length; the passive component is assumed to become null when the elongation becomes negative, i.e. when the muscle length becomes smaller than the rest length.

A third group of isotonic experiments provided evidence that the maximum contractile velocity is nearly independent from the elongation (Gordon *et al.*, 1966; Edman, 1979).

From these observations it possible to reach an analytical formulation of the contractile force:

$$
F_{\mathrm{c}} = \begin{cases} 0 \\[2mm] \dfrac{F(L_0)b + a\dot{w}}{-\dot{w} + b} \\[4mm] \dfrac{3}{2}F(L_0) - \dfrac{1}{2}\dfrac{F(L_0)b' - a'\dot{w}}{\dot{w} + b'} \\[4mm] \dfrac{3}{2}F(L_0) \end{cases} \quad \text{when} \quad \begin{array}{l} \dot{w} < -F(L_0)\dfrac{b}{a} \\[3mm] -F(L_0)\dfrac{b}{a} < \dot{w} \le 0 \\[3mm] 0 < \dot{w} \le -F(L_0)\dfrac{b'}{a'} \\[3mm] F(L_0)\dfrac{b'}{a'} < \dot{w} \end{array} \tag{3.9}
$$

This equation provides a quantitative relationship between the contractile force and the contractile velocity, if the isometric tetanic force $F(L_0)$ is known. Of course, this equation, and similar equations found in the literature, are purely phenomenological models developed using the available experimental data, which always provide an incomplete view of the phenomenon, and which are based on varying levels of idealization.

In addition, the model has some intrinsic weaknesses. The choice of relating the contractile force to the instantaneous velocity and to the length at which the contraction started has relevant implications. Such implications are mostly positive: let us consider the case where the elongation velocity is zero, the muscle is contracted in the position of maximum force, and we try to elongate the muscle further from the outside; in such a case, according to our model, the increase in force would be due only to the elastic stiffness of the muscle. Instead, if we had expressed the contractile force as a function of the actual instantaneous length, in the test described above, the model would have shown an unstable behavior since every additional elongation would decrease the stiffness, further increasing the elongation. However, such a choice of model neglects to take into account that the elastic stiffness itself experimentally shows some dependence on the length at which the contraction started. Another limitation of this model is its inability to incorporate the memory effects that are always observed experimentally, where a sequence of lengthenings, shortenings, and contractions produces forces whose magnitude depends on the order in which they are applied. The limitations of this model somehow reflect also the paucity of experimental observations on the phenomena here described, which would make it difficult to identify more sophisticated models.

3.3.2.5   Muscle contraction dynamics

Even with these limitations the contraction dynamics model described so far is effective in solving all problems that involve a single fiber. If we want to model a whole muscle (which is our case) we need to account for three additional aspects: muscle architecture, tendons, and the level of activation. We can assume that the forces produced by muscle fibers placed in parallel sum up, as do those generated by the various sarcomeres within a single fiber. Such forces point along the direction of the muscle fibers and they are oriented with respect to the muscle's line of action at an angle that depends on the muscle architecture, and the average pennation angle α^{M}. Therefore the mechanical effect of all the muscle fibers at the muscle ends (F^{M}) is the force component along the muscle line of action:

$$F^{\mathrm{M}} = \cos(\alpha^{\mathrm{M}}) \sum_{1}^{n} F_i, \qquad (3.10)$$

where i is the muscle fiber index, which ranges between one and the number of parallel muscle fibers (n). The variable is not defined elsewhere. Here as well as in the following equations, the force exerted by the whole muscle is indicated with the superscript M.

The tendon can be considered by adding another elastic element in series with the rest of the model. Usually, the elastic energy stored in the tendon is so large compared with that stored in the muscle, that when the tendon is included in the model, the viscous components can be neglected. Re-writing Eq. (3.7) to consider the complete musculotendon unit, we have:

$$F^{\mathrm{M}} = F^{\mathrm{T}} = k_{\mathrm{t}} \Delta L^{\mathrm{T}} = \cos(\alpha^{\mathrm{M}}) \cdot (k_{\mathrm{p}}^{\mathrm{M}} u + k_{\mathrm{a}}^{\mathrm{M}} v), \qquad (3.11)$$

where F^{M} is the total muscle force, which is equal to the tendon force F^{T}, k_{t} is the stiffness of the tendon, ΔL^{T} is the variation in length of the tendon unit.

Another aspect neglected so far is the level of activation. The vast majority of available experimental observations used to develop the model here proposed considered only conditions where the muscle is either passive or active in tetanic conditions. In reality muscles rarely reach tetanic conditions during normal motor activities; unfortunately, experimental investigation of the biomechanical behavior of muscles in sub-tetanic conditions is still limited. One assumption that many authors have adopted is that it is possible to express the level of activation with a dimensionless parameter $a(t)$ ranging between zero and one, and that the force–velocity and force–elongation curves scale linearly with such a parameter. Of course, this applies only to the active component of the muscle force, since the passive component is assumed to be totally independent of the level of activation. The equation of the active element is:

$$F_{\mathrm{a}}^{\mathrm{M}} = k_{\mathrm{a}}^{\mathrm{M}} \Delta v = F_{\mathrm{c}}(a(t), \dot{w}, L_0^{\mathrm{M}}, F_0^{\mathrm{M}}). \qquad (3.12)$$

Putting all the equations together, the mechanical effect of a muscle along its line of action can be described by the following system of equations:

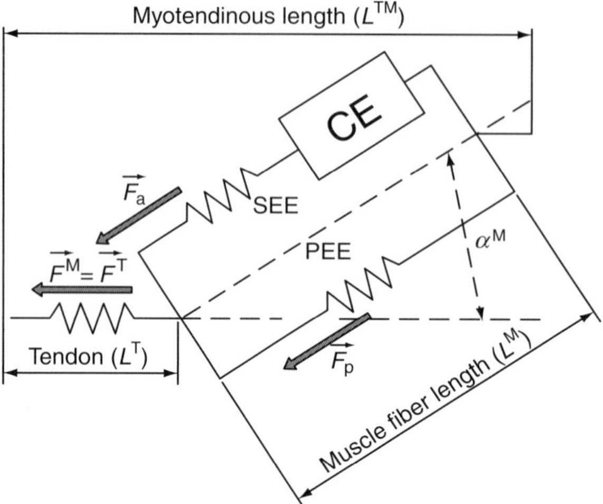

Figure 3.6   Each musculotendon actuator is represented as a three-element muscle in series with an elastic tendon. The mechanical behavior of muscle is described by a Hill-type contractile element (CE) that models the muscle's force–length–velocity properties, a series-elastic element (SEE) that models the muscle's active stiffness, and a parallel-elastic element (PEE) that models the muscle's passive stiffness. The instantaneous length of the actuator is determined by the length of the muscle, the length of the tendon, and the pennation angle of the muscle. In this model, the width of the muscle is assumed to remain constant as muscle length changes.

$$\begin{cases} F^M = k_t \Delta L^T = \cos(\alpha^M) \cdot (F_p^M + F_a^M) \\ F_p^M = k_p^M \Delta u \\ F_a^M = k_a^M \Delta v = F_c(a(t), \dot{w}, L_0^M, F_0^M) \end{cases} \tag{3.13}$$

The graphical representation of this model is shown in Figure 3.6.

The elastic properties and the active force-producing properties are usually expressed through normalized equations that must be identified for each muscle bundle to run a musculoskeletal simulation. In general, several normalized functions have been proposed in the literature and the number and type of parameters depends on the specific formulation. However, some of these parameters are common to all models and have a particular importance in defining the muscle's force-producing properties: the peak isometric muscle force (F_0^M), the corresponding optimal muscle fiber length (L_0^M), the pennation angle (α^M), the tendon slack length (L_s^T) and the intrinsic maximal contraction velocity of the muscle ($\dot{w}_{max}^M$).

3.3.3   Identification of the musculoskeletal dynamics models

So far, I have described whole-body modeling strategies assuming implicitly that all patient-specific information is available. In practice this is rarely true; technical, ethical, and economical limitations always prevent us from obtaining some essential information. In this section, I shall describe how and when it is possible to quantify the information required by whole-body models for a given patient.

To generate a patient-specific musculoskeletal dynamics model, we need to define the skeletal articulated system and the muscular system. For this purpose, we need to identify the intra-segmental local frames, the number and the type of ideal joints and their parameters, the intra-segmental inertial properties, the muscle lines of action, and their functional parameters.

3.3.3.1 Bone geometry

The best way to define the intra-segmental local frames is to collect the patient-specific skeletal geometry. To create an accurate musculoskeletal dynamics model it is necessary to have an accurate 3D geometry of the patient's skeleton. Bone geometry can be derived with good accuracy from computed tomography (CT); however, reasonably good 3D reconstruction of bone geometry has also been obtained using magnetic resonance imaging (MRI) (Lee *et al.*, 2008; Schmid and Magnenat-Thalmann, 2008). Computed tomography uses ionizing radiations, which exposes the subject to some risks; MRI machines are sometimes less available, slower, and usually more expensive. Sometimes the research question emerges from a specific context where ethical, logistic, or economic constraints make one of these imaging methods unacceptable. If this is not the case, then one should choose the imaging method by considering the rest of the modeling that is to be undertaken. Computed tomography should be preferred when an organ-level bone model is also required; as we shall see in Chapter 4, CT data can be used not only to derive the bone geometry but also to estimate the heterogeneous mechanical properties of bone tissue. On the other hand, MRI provides much better imaging of soft tissues, and in particular of muscles; if we need accurate measurements on muscle cross-sections MRI is clearly a better option. Other methodologies, based on 2D imaging, require much more sophisticated methods to derive the 3D bone geometry, and never achieve the level of accuracy we can get with a 3D imaging method; however, in some specific clinical applications they might be the only ethically viable option. (See the application of multiscale modeling to screening for patients at risk of osteoporotic fracture in Chapter 7.)

Special attention should be paid to the imaging method. If CT is used, a helical scan should be preferred as it minimizes the acquisition time, and when cleverly used, also the dose to the patient. To obtain the best spatial reconstruction, the helical pitch should be chosen coherently with the desired reconstruction spacing. If the pitch is too large, slice reconstruction will be difficult, and the image will be fairly noisy. If the pitch is too small, the tails of the X-ray beam irradiation profile of two subsequent spires will overlap, causing a significant increase in the effective radiation dose to the patient. If the CT imaging is done only for modeling purposes, the tube voltage and the exposure time should be chosen considering the unusual purpose of the imaging. Radiologists set these parameters in order to achieve the image contrast required for accurate diagnosis; but image segmentation methods tend to be robust to lower contrasts, and thus if the purpose is modeling lower voltage and lower time can be used (Lattanzi *et al.*, 2004; Van Sint Jan *et al.*, 2006). Lowering these two settings can drastically reduce the effective radiation dose.

Both CT and MRI provide spatial metrics associated with the images that can be used to determine the true dimensions of the modeled bones. Other types of calibration might be necessary, depending on the specific modeling strategy. If the CT scan is made so that an organ-level model can also be built, a densitometric calibration is usually required. If the patient-specific anatomical data must be fused with patient-specific movement data, some segmental registration frame must be provided. This has been achieved by attaching to the patient's skin special markers that are reflective but also opaque under CT or MRI imaging. The markers' positions are reconstructed in the images and also in the movement analysis setting, and provide the fiducial points required to create a segmented reference framework between the anatomical and functional datasets (Leardini *et al.*, 2007).

Once the CT or the MRI datasets are available, they must be segmented to extract from them the 3D geometry of the bones. Various commercial software packages specialize in medical imaging segmentation; the most popular are probably Amira (Visage Imaging GmbH, Germany) and Mimics (Materialise, Belgium). If one has some programming skills, an open-source segmentation library called the Insight Segmentation and Registration Toolkit (ITK)[1] contains some of the most advanced segmentation algorithms. Segmenting bone in an MRI image is usually more challenging, but CT scan segmentation is not so trivial when you deal with real patients, affected by conditions that complicate the procedure. For example, subjects with osteoarthritis (common among the elderly) tend to have reduced joint spaces, making it difficult to separate the bones forming an articulation. Because of this, segmentation generally requires some manual intervention, and can never be entirely automated. In some cases, the manual effort is considerable, and the whole operation may require a significant amount of time.

In principle, to build a complete model one should obtain a full body scan; however, this is usually impractical, too expensive, and sometimes unethical. The most common method is to image only the region of most interest (the hip, the lower limbs, etc.), and then use a generic skeleton properly scaled to the patient's anthropometry to complete the model. Complete 3D skeletal geometries can be found as part of various available collections, including the 3D skeleton of the visible human project (Ackerman, 1998), the VAKHUM project,[2] and the two female skeleton models created during the LHDL project, which are available through the PhysiomeSpace service.[3]

3.3.3.2   Intra-segmental reference frames

Once the 3D skeleton of the patient is obtained, usually as a combination of segmentation of patient CT images and registration of generic skeletal atlases, we must identify the relevant skeletal landmarks on the 3D skeleton. This is called *virtual palpation*, in analogy with movement analysis, where the skeletal landmark positions are

---

[1] www.itk.org/
[2] www.ulb.ac.be/project/vakhum/
[3] www.physiomespace.com/

located by palpating them through the subject's skin. Entire textbooks are devoted to skeletal landmarking (Van Sint Jan, 2007), and there is even a freeware application dedicated to this operation (V-Palp virtual palpation software – SCS, Italy).[4] While the specific choice of landmark depends on the particular bone, the specific modeling purpose, and strategy, etc., it is usually wise to locate as many anatomical landmarks as possible, as they often come in handy in subsequent steps. In particular, it is possible to define intra-segmental local frames from these landmarks. In theory, such local reference frames could be defined in a totally arbitrary way, but in practice it is usually convenient to use anatomy-based standardized reference frames, to simplify the comparison of results between published studies (Wu *et al.*, 2002; 2005).

3.3.3.3   Idealized joints

Once the bone geometry is defined, we need to define the joints linking them. The choice of appropriate idealization for each articulation is a delicate one; however, much of the debate found in the literature on this topic may not be relevant here. Our model will not have to be used to analyze the gait of the patient, or to predict the distribution of the contact pressure in the joint cartilage; so the right way to look at this problem is to ask, "What is the maximum level of idealization I can afford without compromising the accuracy of my model in predicting what I care for (which in my case is the muscle force)?" In general, the large movements allowed along the unresisted directions affect muscle forces much more than those allowed along the resisted directions. Thus, using a ball-and-socket joint at the hip and a cylindrical hinge at the knee might be too crude an idealization in many other problems, but might be adequate in our case. However, before we make this decision, some sensitivity analysis might be worthwhile. The muscle forces can be predicted many times under the same conditions, but changing the position of two bones in the direction of one of the degrees of freedom is suppressed. If our idealization choices are correct, these perturbations should change our prediction of the muscle force only slightly; if this is not the case, the idealization must be revised. One possibility is to permit some of the restricted directions of motion, and then to introduce non-linear springs representing certain ligaments in the model to limit the excursion of the joint along those directions.

3.3.3.4   Inertial properties

Before we can solve our musculoskeletal dynamics model we also need to determine the mass matrix of the patient's body. The mass of the body is usually distributed between the segments of the kinematics chain, lumped into a point mass placed at the center of gravity of the segment. Average body segment inertial parameters are available as a function of body height and weight, tabulated from large population studies (Drillis *et al.*, 1964; Durkin and Dowling, 2003).[5] More accurate information can be obtained using dual X-ray absorptiometry (DXA) (Durkin *et al.*, 2002).

---

[4] www.biomedtown.org/biomed_town/B3C_Building/products/VPalp/
[5] See also www.dh.aist.go.jp/bodyDB/m/index-e.html.

If MRI or CT are available, accurate 3D reconstruction of the outer skin surface is usually possible, and the inertial parameters can be computed by estimating the density of the various tissue types (Pavol, 2002).

3.3.3.5   Excitation–contraction dynamics parameters

To model the contractile dynamics of muscles properly we need to quantify, for each subject, the peak isometric muscle force F_0^M, the corresponding optimal muscle fiber length L_0^M, the pennation angle α^M, the tendon slack length L_s^T and the intrinsic maximal contraction velocity of the muscle $\dot{w}_{\max}^M$

The peak isometric muscle force F_0^M is generally related to its physiological cross-sectional area (PCSA), which is proportional to the number of muscle fibers in parallel. In cadaveric studies, the PCSA is often obtained by dividing the muscle volume by the optimal fiber length (Albracht *et al.*, 2008; Lieber and Fridén, 2000). The muscle volume is usually defined as the ratio between the muscle weight and its average density, typically 1.06 g cm^{-3} (Mendez and Keys, 1960). The PCSA has been tabulated in population studies (e.g., see the appendix of (Yamaguchi, 2001)). Data from MRI are the ideal source for defining muscle volume, but recent software developments also make it possible to compute accurate muscle volumes from CT data (Krokos *et al.*, 2005).

When the muscle PCSA is known, the peak isometric muscle force F_0^M is usually obtained by multiplying it with the tetanic muscle stress, a parameter that reports the maximum force that a unitary area of muscle cross-section is capable of expressing. The underlying hypothesis is that this parameter is a sort of inherent property of muscles, and should not change much between muscles or between individuals. However, it has been noted (Buchanan *et al.*, 2004) that the range of values reported in the literature (0.35–1.37 MPa) is too wide to inspire confidence. It is unclear whether there are methodological problems in the experiments used to quantify it, or whether there is a considerable objective intra- and inter-subject variability for this parameter, which would nullify its usefulness.

Values of the optimal muscle fiber length L_0^M , the pennation angle α^M, and the tendon slack length L_s^T are almost always based on population data obtained from cadaver dissections (e.g. (Yamaguchi, 1990)). The average pennation angle can be measured using imaging methods (MRI or ultrasound), both on cadavers and on single patients. However, there is no method for measuring the optimal fiber length directly, since it is not clear whether there is a single posture whose associated fiber length can be assumed to be optimal. However, it has been observed that the optimal length of the sarcomere varies within a very narrow band (2.7–2.9 microns). Thus, during dissection studies, we can measure the actual length of a muscle fiber, its actual sarcomere length, and then derive the optimal fiber length as a proportion with the ratio at the sarcomere level. Also the tendon slack length can, in theory, be visualized with functional imaging, but in practice for the majority of the muscles this is very difficult, owing to the anatomo-functional complexity of the musculoskeletal system. Since the tendon is frequently represented by an elastic element, it is possible to assign to it a nominal length that is required by the modeling constraints, which can be estimated numerically (Winby *et al.*, 2008). This works as far as we can assume that the tendon has a linear elastic behavior, which means neglecting the initial non-linear "toe" region of the curve. This simplification overestimates the amount of strain energy stored in the tendon, but the effect on the performance of the muscle actuator is unlikely to be significant because the tendon force is small in the region where the force–length curve is non-linear. Still, the ratio between the optimal muscle fiber length L_0^M, and the tendon slack length L_s^T has a significant effect on the predicted muscle biomechanics. The intrinsic maximal contraction velocity of the muscle $\dot{w}_{max}^M$ is different for fast and slow fibers (Zajac, 1989). $\dot{w}_{max}^M$ is assumed to be muscle-independent in most simulations of movement and it is given a conventional value, i.e. $\dot{w}_{max}^M = 10\,L_0^M$.

To date, the level of knowledge we have on the effect that each of these parameters have on the predictions of a musculoskeletal dynamics model is surprisingly small. Also, it is unclear what error can be expected when using a population-based value instead of a subject-specific value for each of these parameters. Further studies are clearly needed in this area.

3.3.3.6 Lines of action and lever arms

The next operation required is the definition of the myotendinous actuators' action lines. As explained in the introduction, we assume that the muscle–tendon complexes can be idealized as bundles of rectilinear actuators, each connected to only two bones. The representation of the various muscles using a handful of linear actuators requires a very careful choice of the number, location, and orientation of these actuators. Apparently, the operation is very simple. Each muscle bundle has an origin and an insertion; we compute the center of the insertion area and of the origin area, and then we define the linear actuator and the connection between these two centers. In practice, this works only for a few muscles, because of a few additional issues. Not all muscles have a single origin or a single insertion; anatomists usually define the capita of the muscle, and from these we can split the muscle volume into two or more separate bundles, each with a single origin–insertion pair. However, this decomposition is not trivial, and very much depends on the individual patient anatomy, given the considerable inter-subject variability of these anatomic features. Some muscles have complex pennation patterns (i.e., the muscle fibers within the muscle volume are oriented in complex ways); in some cases, it is more convenient to split a muscle into two or more bundles, so that for each bundle the pennation angle (the angle between the average fiber orientation and the muscle line of action) is reasonably constant. Another issue is muscle wrapping. Many muscles do not follow a straight line, but rather wrap around a bone; of course wrapping is not static, during motion muscles can wrap and unwrap as their position with respect to the underlying bones changes. Currently no musculoskeletal modeling program can automatically detect the wrapping of the muscle lines of action onto the bone surfaces (although various codes can compute the wrap around user-defined parametric surfaces that locally approximate the bone surface), and in a way this would not be desirable: in reality the wrapping is the result of a complex interaction between the skeleton, the various muscles, other connective tissues such as fascia, and the adipose layers. Thus it is necessary to recognize a priori whether a muscle will eventually wrap during the movement, and in such cases to identify the wrap point or the wrap region, i.e. the location on the bone surface where the muscle line of action bends around the bone surface. Because of all these issues, the definition of the systems of linear actuators that model the muscles is a complex operation, not much standardized, and requiring a lot of experience. This is why many authors suggest using a generic model, and adapting it to the patient-specific skeleton.

The simplest approach is to determine the affine transformation (roto-translation plus anisotropic scaling) that minimizes the distance between the sub-set of skeletal landmarks that are present in both the atlas model and in the patient-specific model. This transformation is then also applied to the landmarks that mark the origin, insertion, and, eventually, wrap point of the linear actuators defined in the generic model. In general, this transformation places the landmarks close to but not exactly onto the bone surface, so some manual adjustment is required. In any case, this operation provides only a first approximation to the patient's muscular anatomy. Using data fusion environments, such as Data Manager (Viceconti *et al.*, 2007), the resulting musculoskeletal model should be superimposed on the patient CT or MRI images, to verify that the linear actuators are correctly placed with respect to the patient-specific anatomy, and eventually make fine adjustments.

A second approach involves the segmentation of muscle volumes in the patient's images. If MRI data are available, segmentation is relatively simple, although time-consuming (Gilles and Magnenat-Thalmann, 2010). However, if only CT images are available, muscles shapes usually appear with very low contrast, and direct segmentation is very difficult. In these cases, it is possible to use template-based elastic registration, where a generic 3D muscle model is adapted to the patient anatomy interactively using the CT image as guide (Krokos *et al.*, 2005). One way or the other, when the 3D shapes of the muscles are available they can be used as a guide to define the line of action of the linear actuators that model them.

It should be mentioned that once we have the muscle 3D geometry, in theory we could build an organ-level model (Blemker *et al.*, 2005; Fernandez *et al.*, 2005). Using finite-element analysis it is possible to model both the passive (Calvo *et al.*, 2010) and the active (Rohrle and Pullan, 2007) behavior of skeletal muscles. Thus, in theory one could make a whole-body model where each bone and each muscle is modeled as a deformable solid using finite-element analysis. However, apart from the fact that such a model would be computationally too heavy to be solved with most available computers, such a modeling strategy would burden us with a lot of unnecessary complexity. Our goal is to predict the stresses and strains of bones during movement; we are not interested in knowing the mechanical strain in the muscles, but only the force they exert on bones during contraction. Generality can be a useful feature of the software tools we use to pre-process the patient information and convert them into a predictive model; but the model itself should always be tailored to the specific research question we aim to answer.

Once we have defined the muscles' lines of action, we need to compute the lever arm of each muscle with respect to each joint it crosses. Lever arms are classically defined as the distance between the line of action and the joint center. A more general definition of moment arms is obtained using the principle of virtual work (An *et al.*, 1984). If $R_{ij}(q)$ is the moment arm of muscle j with respect to joint axis i, we can write:

$$R_{ij}(q) = -\frac{\partial L_j(q)}{\partial q_i}, \qquad (3.14)$$

where $L_j(q)$ is the origin–insertion length of muscle j as a function of all joint degrees of freedom q. It should be noted that, in principle, this approach enables the use of curvilinear lines of actions for modeling muscles.

3.3.3.7 Skeletal kinematics and the ground reaction

Before the body inverse dynamics model can be solved, we need to quantify its inputs, i.e. the skeletal kinematics, and the external forces acting on the body during the motion, which are normally represented by the ground reaction.

The definition of the skeletal kinematics is a fundamental step in predicting how the muscles contribute to movements on a specific subject. Current techniques are mostly based on traditional infrared optoelectronic stereophotogrammetry methods and only recently have more economic, full 3D motion-tracking systems based on wearable wireless inertial sensors appeared on the market. However, all the common techniques that can be used in recording the skeletal kinematics of the full body in a convenient manner are hindered by the impossibility of measuring the instantaneous position in space of skeletal segments directly. All available methods record the kinematics of points attached to the skin, and not on the underlying bones; the problem is solved either by assuming that those points on the skin are rigidly attached to certain anatomical landmarks on the bones, or by assuming, a priori, a map of movements of the skin relative to the bones during motion (Leardini *et al.*, 2005). Under these assumptions, the motion-tracking trajectories of the skin points are transformed into trajectories of relevant bone landmarks, which are then used to compute the trajectories in space of the intra-segmental anatomical frames (Cappozzo *et al.*, 2005).

Different categories of error can affect the kinematics predicted using these methods: errors introduced by the instrumentation's accuracy (generally low), errors resulting from the assumption of a rigid connection between skin and bones (or an a-priori defined map of skin movements) and errors made in identifying the position of anatomical landmarks, resulting in the incorrect position of anatomical frames. The most relevant are the errors from skin–bone relative movements (known as *soft tissue artifacts*). Several methodologies have been proposed to compensate for this type of error. In particular, global optimization and joint constraints were found to be effective in the reduction of the effect of skin artifacts on kinematics predictions globally (Lu and O'Connor, 1999).

To measure the ground reaction forces, the most common approach is to embed a *force platform* in the floor where the motion capture is performed. The most sophisticated force platforms can measure all six components of the generalized force vector with excellent accuracy, if properly calibrated. The biggest limitation of this approach is that the sensing area is fixed and relatively small, so that it can only capture a small portion of the motion. In some settings, multiple force platforms are used to reduce this problem. The other option is, again, to use wearable sensors, such as pressure-sensitive insoles, but their accuracy in measuring the resultant of the pressure distribution (ground reaction) still appears to be insufficient (Hsiao *et al.*, 2002).

3.3.4   Predicting muscle forces by inverse dynamics and constrained optimization

If the musculoskeletal anatomy and the segmental kinematics are known, inverse dynamics can be used to compute the joint moments M_n. To ensure instantaneous equilibrium the following condition must be satisfied:

$$M_n = \underline{\underline{R}}(\underline{q}) \times \underline{F}^{\mathrm{M}}, \tag{3.15}$$

where M_n is the vector containing the resultant moments at all the n joints, $R(q)$ is the matrix of the muscular moment arms ($n \times m$) and F^{M} is the vector of the myotendinous forces ($m \times 1$) expressed by the m independent muscle bundles considered in the model. Since $m > n$, there is an infinite number of muscle force vectors that equilibrate the joint moments in every instant.

The muscular load-sharing problem can be solved for each instant in time, by minimizing an objective function J subject to constraints representing the equality of the sum of individual muscular moments to the joint torques calculated from the inverse dynamics analysis. The individual muscular moment is calculated from the muscle force (the unknown of the optimization problem) and muscle moment arms, which are derived from the musculoskeletal anatomy and may or may not depend on joint angles. The maximum possible force that each muscle can express is limited and this introduces an additional constraint. Thus, the solution is obtained by solving the following constrained minimization problem:

$$\begin{aligned}
\text{minimize} &\rightarrow J\left(\underline{F}^{\mathrm{M}}\right) \\
\text{subject} &\rightarrow \begin{cases} M_n = \underline{\underline{R}}(\underline{q}) \times \underline{F}^{\mathrm{M}} \\ 0 \leq \underline{F}^{\mathrm{MT}} \leq \underline{F}_{\max} \end{cases},
\end{aligned} \tag{3.16}$$

where $F_{\max}$ is the vector of maximum forces that each muscle can express. This upper limit can be computed dynamically by incorporating the equations that model the musculotendon dynamics in an implicit form.

Since the seminal work of Seireg and Arvikar in the mid seventies (Seireg and Arvikar, 1975), there has been an extensive debate in the literature over the form of the optimization function. However, since it is not possible to measure muscle forces non-invasively in vivo, the discussion has been somewhat inconclusive. In some studies (Collins, 1995; Menegaldo *et al.*, 2006; Praagman *et al.*, 2006; Tsirakos *et al.*, 1997), the minimization of the total muscle stress has been found to produce muscle-force activation patterns in good agreement with EMG recordings. In this case the minimization function J takes the form:

$$J\left(F^{\mathrm{M}}\right) = \sum_i \left(\frac{F^i{}_m}{PCSA^i}\right)^k,$$ (3.17)

i.e. the summation of the muscle tensions computed from the force of each muscle divided by the physiological cross-sectional area (PCSA), raised to an exponent k, which empirically represents the activation strategy, i.e. the synergistic contraction of multiple muscle to produce a certain body movement. If $k = 1$, the solution does not include muscle synergism for small loads, whereas for $k > 1$ the solution includes synergism and sometimes also *antagonism* (a muscle that acts in opposition to the specific movement). When $k \to \infty$ the solution maximizes synergism, thus minimizing muscle fatigue.

3.3.5 A probabilistic approach to daily load histories

While the muscle forces predicted with the methods described in the previous section have been found to be in good agreement with EMG recordings in a number of studies (Erdemir *et al.*, 2007), and the intensity of the hip joint reaction predicted by the models is comparable to that recorded with telemetric instrumented prostheses (Bergmann *et al.*, 2001), the lack of direct measurements prevents us from knowing the precise accuracy of these predictions.

In addition, we should not forget the previously mentioned "uncontrolled manifold" (UCM) theory (Scholz and Schoner, 1999), which suggests that the motor-control strategy focuses on the goal of the task, and that every trajectory within the manifold of the task-equivalent configuration of the muscle actuators is virtually possible. If we look at studies where the same subject repeats the same task a number of times, we shall be impressed by the variability of the kinematics and of the internal and external forces (Bergmann *et al.*, 2001). The approach described in the previous section presumes that the neuromotor control chooses, from among the infinite solutions available, the muscle activation pattern that optimizes a certain cost function, and that it always chooses the same pattern. But this is unrealistic: when we move, sometimes our goal is to maximize the performance, sometimes to minimize the energetic expenditure, other times to minimize the forces acting on a given muscle, or joint, or ligament (perhaps because they are painful). The way we move is also affected by our emotions; if I am happy it is more likely that I shall walk with a faster pace, longer strides, a straighter posture, etc. Depression has been found to be a co-factor in the risk of falling in the elderly (Sai *et al.*, 2010), while somatization, anxiety, and depression were found to be intrinsic co-factors in non-specific musculoskeletal spinal disorders (Manchikanti *et al.*, 2002). Psychological condition was found to significantly affect the so-called six-minute-walk distance (6MWD), a classic geriatric indicator for general musculoskeletal health and mobility (Lord and Menz, 2002).

For all these reasons, in a number of cases the neuromuscular indeterminacy problem should not be solved with a deterministic optimization approach, but rather from a probabilistic perspective. The probabilistic investigation of the neuromusculoskeletal system is a very recent perspective in biomechanics research, and is still an open research topic. Next, I summarize some recent preliminary results obtained by describing the neuromotor control as a non-optimal process, in order to understand if this condition involves significant skeletal overloading.

The analysis of the variability of the hip load magnitude due to sub-optimal neuromotor control strategies is performed considering the frame for which the first peak load at the hip was predicted under the assumption of optimal neuromotor control. The instantaneous equilibrium of the model is achieved by every set of muscle forces that satisfies the following system of equations:

$$\begin{cases} \overline{\overline{B}}_{m,n} \bullet \overline{F}_n = \overline{M}_m \\ F_n \in \left[0, \overline{F}_{\max} \right] \end{cases}, \qquad (3.18)$$

where $\overline{\overline{B}}_{m,n}$ is the matrix of the lever arms, $\overline{F}_n$ is the vector of the muscle forces, $\overline{M}_m$ is the vector of the articular moments, and $\overline{F}_{\max}$ is the vector of the maximum tetanic forces that each muscle can express.

The totality of the solutions is given by the infinite set of points internal to a bounded portion of a hyperplane (i.e. hypersimplex) defined in the unknown hyperspace of possible muscle forces. Preliminary simulations suggested that much larger variations on the hip load were possible by changing the muscle forces along the direction of the hypersimplex connecting the two neuromotor control strategies producing the maximum ($HR_{\max}$) and the minimum ($HR_{\min}$) values of the hip reaction magnitude, respectively (principal force vector hereinafter). Thus, we sampled the principal force vector at 10^3 sampling points and we perturbed each intermediate solution by 10^2 random combinations of the remaining transversal base vectors. The process resulted in 10^5 values of the hip load magnitude produced by each sample of the computed sub-optimal neuromotor control. The differences between the magnitude of the hip load produced by the sub-optimal and the optimal controls were also calculated, and expressed in multiples of body weight (BW). The procedure we used to compute the 10^5 samples of dynamically equivalent neuromotor muscle controls involved a few steps:

- The principal force vector is defined through optimization, computing the neuromotor controls producing the minimum ($HR_{\min}$) and the maximum ($HR_{\max}$) value of the hip reaction magnitude.
- The principal force vector is sampled by 10^3 uniformly distributed sampling points ($\overline{F}_{pa}$).

- An orthonormal base $\bar{v}_n^j$ of the homogeneous system of equations is computed through a singular value decomposition so that

$$\bar{\bar{B}}_{m,n} \bullet \bar{\bar{F}}_n = 0, \tag{3.19}$$

$$\bar{\bar{B}}_{m,n} \bullet \bar{v}_n^j = 0; \qquad \forall j = 1, \dots, n - m, \tag{3.20}$$

where $\bar{v}_n^j$ is the base vector j of the $n - m$ base vectors that span the solution space of the homogeneous system, Eq. (3.20).

- The matrix $\bar{P}^k$ containing 100 random directions on the solution space is computed through linear combinations of the base vectors:

$$\bar{P}^k = \sum_j \alpha_j \bullet \bar{v}^j; \quad j = 1, \dots, n - m; \quad k = 1, \dots, 100, \tag{3.21}$$

where $\bar{P}^k$ is the k-column vector that represents a random direction on the solution space, and α_j is a random coefficient.

- For each sample point on the principal force vector $\bar{\bar{F}}_{\mathrm{pa}}$, two additional points on the boundary of the hypersimplex are computed:

$$\bar{F}_{1,2}^{j,k} = \bar{F}_{\mathrm{pa}}^j \pm \beta_{1,2} \cdot \bar{P}^k, \tag{3.22}$$

where $\beta_{1,2}$ are two coefficients so that the two sample points $\bar{F}_{1,2}^{j,k}$ are motor-control strategies on the boundary of the hypersimplex, both lying in the direction $\bar{P}^k$ through the point $\bar{F}_{\mathrm{pa}}^j$.

- The magnitude of the hip reaction load is calculated for each sub-optimal neuromotor control scheme as:

$$HR_{1,2}^{j,k} = \left\| \overline{HR}_{\mathrm{dyn}} + \sum_{j,k} \bar{F}_{1,2}^{j,k} \right\| / \mathrm{BW}, \tag{3.23}$$

where $HR_{1,2}^{j,k}$ is the normalized magnitude of the hip reaction load, $\overline{HR}_{\mathrm{dyn}}$ is the hip reaction vector calculated through the inverse dynamic simulation and $\bar{F}_{1,2}^{j,k}$ are the muscle forces acting on the hip calculated in Eq. (3.22).

The peak force that a muscle can exert is conventionally estimated through two different parameters: the muscle's physiological cross-sectional area (PCSA) and the tetanic muscle stress (TMS). However, while several methods can be adopted to extract reliable information on the PCSA, estimation of the TMS is more complex and the literature reports a very large range of values (0.35–1.37 MPa). Thus, we repeat the estimation of the hip load boundaries across the entire TMS range.

When this approach was applied to a generic musculoskeletal model, the hip load predicted by the model under the assumption of optimal neuromotor control showed a peak at 17% of the gait cycle. For this frame, the samples of the principal force vector induced variations of the hip-load magnitude from the optimal control solution of up to 3.8 BW. Perturbing the intermediate samples by random combinations of base vectors, the variations on the hip-load magnitude were always below 1 BW. At 17% of the gait cycle, equilibrium at the joints was found for TMS values in the range 0.57–1.37 MPa; when TMS was in the range 0.35–0.57 MPa, the excessive muscle weakness did not guarantee the dynamic balance of the motion. Assuming TMS = 0.57 MPa, the hip load ranged between 3.77 BW and 4.26 BW. This is 13% higher than the lower boundary of the joint load. When the TMS was increased to the maximum value (1.37 MPa) in uniform steps of 0.1 MPa, the lower boundary of the hip-load magnitude did not increase significantly. On the contrary, the variation of the upper boundary of the hip load was significant: for the lowest TMS value the maximum hip-load magnitude was 4.26 BW while for the highest TMS value it was 8.93 BW, 275% higher than the corresponding lower boundary of the joint reaction.

The approach described here presents a number of limitations, and should be considered preliminary. Owing to the considerable computational costs involved, the study was limited to a single instant of the walking cycle. More efficient algorithms are required to extend the calculations to the entire motion cycle, to different motor tasks, and multiple repetitions of the same task. Furthermore, the method we developed to sample the solution hyperspace does not guarantee that the sampling covers all the hypersimplex.

Even if these limitations are solved by improved methods, the approach described here provides us with the solution space of all physically and physiologically possible loading patterns. But this does not mean that all of them will necessarily be observed in real subjects, or that all solutions will occur with the same probability. Indeed, the determination of the probability density function that we need to associate with the hypersimplex of all possible solutions remains an open problem to be addressed in future research.

4   Organ level: the bones

A description of the anatomy and the physiology of bones, and the methods used to model whole-bone biomechanics, in order to predict displacements, stresses, and strains induced by loads acting on the skeleton.

4.1   Elements of anatomy and biomechanics

4.1.1   Bone anatomy

The body of vertebrates is supported by an endoskeleton made of connective tissue in cartilaginous and mineralized form. The latter is organized in rigid organs called *bones*. Bones are organs in the full sense: while we tend to imagine them as inert things, each bone is a true organ made of specialized cellular populations, a complex extracellular matrix, a sophisticated vascular system, and an extensive innervation that is still the subject of investigation.

A first classification of bones is related to their location in the body. Following *Gray's Anatomy* (Gray, 1918), we can distinguish between the *axial skeleton* (vertebral column, thorax, skull) and the *appendicular skeleton* (upper and lower extremities). A third, separate, group is that formed by the *auditory ossicles*. The vertebral column includes the entire spine (cervical, thoracic, and lumbar), the sacrum and the coccyx; the thorax includes the sternum and the ribs; the skull includes all cranial and facial bones, down to the hyoid bone in the throat; the ossicles are the middle ear bones, tiny bony structures that transmit sound inside the ear. The upper extremities can be further divided into the shoulder girdle, the arm, and the hand; the lower extremities into the pelvis, leg, and foot. Overall, a normal adult has 206 distinct bones. A child's skeleton has a much larger number (around 300), but a number of small bones fuse during growth.

A second way to classify bones is according to their shape. The classic Gray's taxonomy includes long, short, flat, and irregular bones. A fifth group, that of sesamoid bones, is included in more recent anatomy texts. Long bones are mostly common in the limbs; they are characterized by a slender shaft, called the *diaphysis*, and two extremities, called *epiphyses*. Short bones are bulk blocks of cancellous bone wrapped by a thin layer of compact bone; examples of short bones are the wrist and ankle bones. Flat bones are made of two layers of compact bone sandwiching a layer of cancellous bone, and usually have a protective function, such as in the cranial bones, the sternum, or the scapulae. We call irregular the bones that do not fit into any of the previous categories; the vertebrae and various facial bones are considered irregular. Sesamoid bones are bones embedded into tendons, such as the patella in the knee.

As already mentioned, there are various types of bone tissue; this argument will be expanded in the next chapter, when we shall focus our attention on the tissue level. However, it is worth stressing here that the difference between compact and cancellous bone is purely histological and, as with all distinctions, the regions of transition are fairly questionable. The whole extracellular matrix of the bone is composed of collagen and hydroxyapatite organized into tightly packed lamellae. In some regions, the bone tissue is so porous that the lamellae form an interconnected three-dimensional network of slender trabeculi, resembling a sponge; in other cases, the porosity is so low that the entire bone volume is filled with lamellae. If we observe the bone tissue from a chemical and a densitometric (i.e., meaning the mineral concentration) point of view, the discrimination between compact and cancellous bone is totally arbitrary. It is only when we observe the tissue morphology, using conventional histomorphology or microtomography, that we can recognize the spatial organization of bone tissue, and thus discriminate regions of compact and regions of cancellous bones; and even in histological observation the demarcation between the two types is extremely blurred, with wide regions where compact bone is highly porous, or cancellous bone is very dense. This important clarification will become useful when discussing the constitutive equations of bone tissue.

4.1.2   Bone biomechanics

Given that the two most important functions of the skeleton are those of protection and support, it is not surprising that an understanding of how bones react to the application of intense mechanical forces is of vital importance.

When we apply a force to a solid body, this body in general both accelerates and deforms. However, it is usual in mechanics to separate these two phenomena, first assuming that the body is infinitely rigid and investigating its kinematics and its dynamics; and then assuming that the system of applied forces is in static equilibrium, so that the velocity and the acceleration of the center of gravity are null, and only the deformation of the body is investigated.

In Chapter 3 we investigated the modeling of the skeleton at the whole-body scale; this is usually the preferential scale for investigating the relative rigid motion of bones. Thus, in this chapter we shall assume that our bones are always subject to a system of forces in equilibrium, so that the acceleration of the center of gravity is null, and only the deformation of the solid body is investigated. To model the biomechanics of bones under this assumption, we rely on the methods of a branch of physics called *continuum mechanics*; in particular, since bones can be considered as solid continua, *solid mechanics*. The definition of a continuum has already been provided in Section 2.3.2.

In solid mechanics, the deformation of a solid body under the action of a system of external forces or displacements can be described using a system of partial differential equations (PDEs). Partial differential equations can be solved numerically in a number of ways, but when the boundary that delimits the domain of integration (i.e., the external surface of the body) is of complex geometry, and external forces act on it, the PDE is called a boundary-value problem, and in this case the method of choice is the *finite-element method* (FEM) described in Chapter 2.

A finite-element model is entirely defined when we can detail the geometry of the solid body, its discretization into finite elements, the constitutive equations that describe how the materials composing the body behave under the action of mechanical forces, and the geometry of the boundary and the conditions that are imposed on it. If the model is to be used to conduct a failure analysis, to this list we must also add the failure criterion that can be used to define the external conditions under which the body fails. Last, but not least, as for any model we need to ensure that the predictions of the model are sufficiently accurate, and this requires that we define verification and validation criteria.

The rest of the chapter will be constructed around these elements. I shall discuss each in turn, showing when and how it is possible to detail these elements using information relative to a specific subject, so that we can generate a bone biomechanics model that is subject-specific in its predictions.

4.1.3   Source of modeling information

When we wish to develop a model of a given bone, it is important to know what information is available on that bone, and the source of that information. It is evident that it is quite different if we need to model a cadaver bone, on which we can make all possible non-destructive and destructive measurements, or if we need to model a bone in vivo. Still, as we shall see, a very good source of geometric and biomechanical information on bones, computed tomography (CT), can be used both in vivo and ex vivo on cadaver bones. This method is so effective that in the last few years almost all modeling studies have relied on it as a primary source of information.

Because of this, in the following I shall always presume that it is possible to have a detailed CT scan of the bone to be modeled, except in the section on geometric modeling, where I shall mention other data sources for completeness.

4.2   Modeling bone geometry

4.2.1   Definitions

The outer surface of bones is delimited by a dense connective tissue called the *periosteum*, which wraps any part of the bone surface that is not covered by articular cartilage. With a few exceptions (such as sesamoid bones, which are embedded into tendons) the separation between the bone tissue and the surrounding soft connective tissues is quite clear, and the concept of bone geometry is somehow meaningful.

The outer layer of the periosteum is made of fibrous tissue dispersed with fibroblasts, whereas the internal layer, called the *cambium layer*, is populated by progenitor cells that can differentiate into chondroblasts or osteoblasts, i.e., cartilage- or bone-forming cells. The other connective tissues (tendons, fasciae) all attach to the bone through the periosteum. This is possible thanks to strong bundles of collagen fibers, called *Sharpey's fibers*, which penetrate the periosteum and the superficial lamellae of bones, ensuring a solid attachment between the periosteum and the other connective tissues. This makes it extremely difficult to separate the ligament or the tendon from its bony insertion, as the penetrating fibers blend into the bone tissue microstructure.

A second geometry, much harder to be defined precisely, is that delimited by the *endosteum*. This layer of connective tissue is present only in bones with a medullary cavity (mostly long bones), where it covers the wall of the medullary cavity, separating the marrow from the bone.

Even harder to identify precisely is the geometric distinction between bone tissue types. In general, there are patterns related to the bones' shapes. For example, in long bones, diaphyses are mostly made of osteonal cortical bone, whereas epiphyses are mostly made of cancellous bone covered by a layer of lamellar cortical bone. While some authors describe the separation between cancellous and cortical bone as another geometrical feature of bones, such a distinction is quite blurred as one tissue type blends into another over a quite broad transition region; in addition, the typing is an histological property, which does not makes much sense at the organ level, where we assume that the bone tissue as a continuum. Thus in the rest of this chapter, such distinction between tissue types will be neglected, and such differences will be accounted for only in the constitutive equations.

A good question is what could be the level of accuracy that we need when defining the bone geometry for modeling purposes. In a sensitivity-analysis study, Taddei and co-workers (Taddei *et al.*, 2006) found that for stress predictions, which are the predictions most affected by geometric uncertainties, the coefficient of variation was 4% or less. This resulted from various uncertainties, but the highest coefficient of correlation for stresses came from the geometric uncertainty. In this study, the geometric uncertainty was represented with a scaling factor applied to the nominal shape; this factor was defined as a normally distributed random variable, with an average value equal to one, and the distribution was truncated at six standard deviations. This latter value of scaling produced a peak deviation of the femoral geometry from the nominal value of slightly less than 10 mm, in the greater trochanter region. In this case, the error on the stresses produced by this single error was around 10%. For an accuracy in stress prediction of 95% or better, which is necessary in a number of concrete problems, this means that we shall have to reconstruct the bone geometry with an accuracy better than one millimeter on average, and better than five millimeters as peak error (not located in regions of high stress).

Over the years, researchers have used disparate methods to reconstruct the three-dimensional geometry of bones. Each method has its pros and its cons, and remains more or less adequate depending on the scope of the modeling exercise. Next, I will review the most commonly used methods, discussing each one in depth.

4.2.2 Serial sectioning

Being mineralized, bones maintain their shape after being dissected from cadavers, and thus it is possible to reconstruct their geometry by serially sectioning them, digitizing the contours of the slices, and then interpolating them to generate the three-dimensional geometry. For small bones it is possible to rely on the methods used for bone histology, but for long bones it is necessary to devise custom-made jigs that ensure the alignment of the slices over the entire bone length. Such an approach to the reconstruction of bone geometry is complex, time-consuming, and destructive for the cadaver bone, and has now almost completely been abandoned. It might become useful if one needs to identify complex constitutive equations at the tissue level, where the quantity, location, and orientation of specific components of the extracellular matrix are accounted for in the constitutive equation; but this is an aspect I shall deal with in Chapter 5.

4.2.3 Digitizing cadaver bones

If our source of information is cadaver bone, and if the only geometric information we want to recover is the outer bone surface, digitizers provide much greater accuracy without destroying the bone. A number of technologies can be used to locate points in the bone surface, and the three-dimensional geometry can be reconstructed from these. They vary in complexity, accuracy, cost, and requirements in terms of surface preparation. Three-dimensional contact digitizers can be as precise as 0.01 mm in locating a point in space, with volume measurement more than adequate for the largest bones, such as the femur. Optical methods, based on coherent light scanners, can capture thousands of points in seconds, although in many cases they require some pre-conditioning of the surface. Still, there are a number of factors that lessen the accuracy of the 3D reconstruction even if the inherent accuracy of the instrumentation is so good. In general, the higher the accuracy, the longer the time required to prepare the specimen and to conduct the measurements. While in some cases the method can be used as an alternative source of geometric information in methodological studies, in most studies reconstruction from tomographic images is the preferred method.

4.2.4 Tomographic imaging

Computed tomography (CT) measures the attenuation of an X-ray beam that is projected in multiple directions radially through the body being imaged. These attenuations are then used to compute the coefficients of attenuation in each point of the body lying in the plane defined by the irradiation circle. By moving along the axis normal to such a plane, a CT scanner can create a "stack of slices" that can be

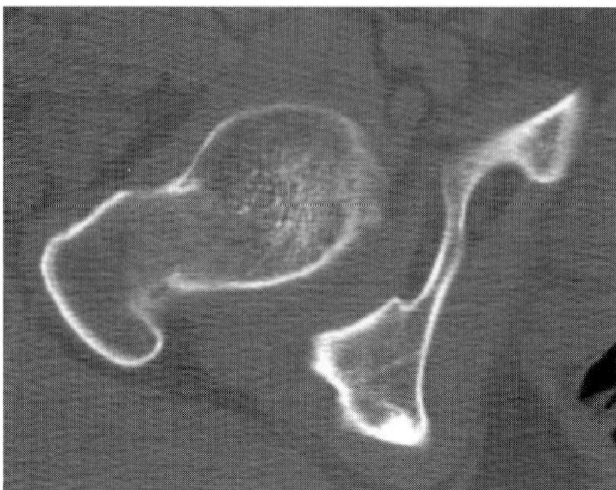

Figure 4.1   A CT slice in the proximal femoral region: the good contrast between the mineralized bone tissue and the surrounding soft tissue is evident.

interpreted as a 3D image of the body's attenuation to X-rays. In modern machines, the acquisition along the circle and the axial translation happen simultaneously (helical or spiral CT), which speeds up the process and in most cases reduces the radiation dose to the patient. The attenuation is mapped in CT images using a scale called *Hounsfield units*, where water has an attenuation of zero Hounsfield units. For a detailed description of computed tomography see (Kalender, 2005).

X-ray attenuation depends on a number of factors, but in general the higher the density of the tissue, the higher the coefficient of attenuation. Since bone is the densest of all body tissue, it is not surprising that bones show with very good contrast in CT images, and can in general be well separated from the surrounding soft tissues (Figure 4.1).

It is thus possible to process each CT image digitally so as to separate bone pixels (which is called *classification*) and then to compute the closed planar curve that separates the bone from the rest of the body (which is called *segmentation*). This stack of planar curves can then be used to reconstruct the 3D geometry of the bone (Figure 4.2).

Although the process sounds quite simple, there are many limitations that can drastically reduce the accuracy of the data obtained. A detailed discussion of this procedure is beyond the scope of this book, but because of its importance the most relevant issues are briefly described, together with some references to methodological papers for those who need greater detail.

The first problem one usually faces is that in the hospitals where CT scanners are usually available these machines are not used to reconstruct geometries but rather to make diagnoses. The difference is more important than one might at first imagine. In diagnosis it is of vital importance that the entire region of interest is scanned, to ensure that every clinically relevant sign appears in the images. This is done by using fairly large "slice thicknesses"; as a result, the tomographic image shows an average

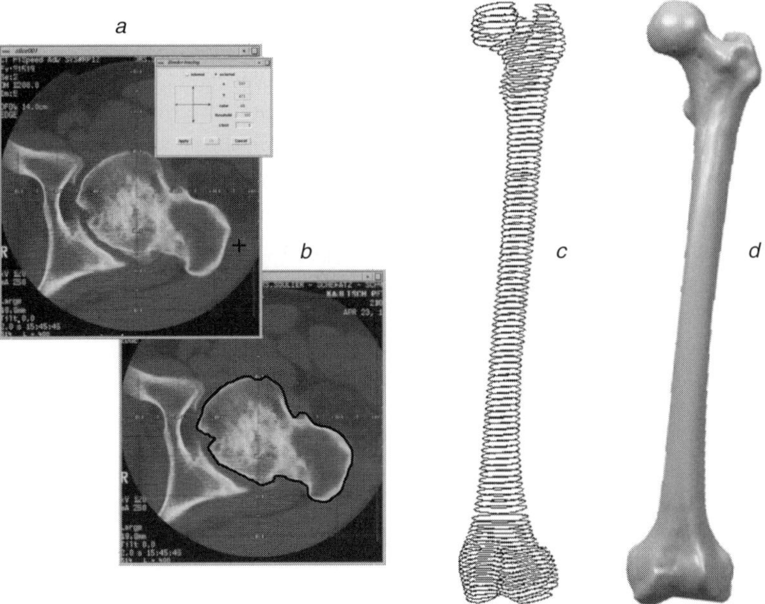

Figure 4.2 The segmentation of a CT dataset of a femur provides a closed planar curve for each slice of the dataset (b). The stack of curves (c) can be used to reconstruct the 3D geometry of the bone (d).

attenuation over this thickness. However, if our goal is to reconstruct bone geometry, thick slices can produce considerable geometric error, especially in regions where there is a considerable radial gradient in the geometry along the tomographic axis. Thus, in spite of recommendations of trained radiologists, one must request the thinnest possible slice. With conventional machines, the spacing of the slices along the tomographic axis is of vital importance; ideally it should be defined on the basis of the aforementioned geometric gradient (Zannoni *et al.*, 1998). With helical CT, what matters is the ratio between the so-called "pitch" of the irradiation spiral and the actual spacing of the reconstructed slices. Even in this case, it is recommended that in long bones, such as the femur, at least two separate pitch or spacing settings be used for the epiphyses and the diaphysis (Lattanzi *et al.*, 2004).

In general, there is a trade-off between the amount of ionizing energy delivered to the body and the quality of images that can be obtained. If dose is not an issue (as with cadaveric bones), one should use relatively high-intensity settings. However, when the dose is a concern, imaging for modeling is usually less demanding than other diagnostic usages, and relatively low intensities are allowed (Van Sint Jan *et al.*, 2006).

Another important parameter is the so-called *field of view* (FOV). This defines the circular region within the machine gantry that will be imaged. The final image is sampled over a fixed-resolution matrix, typically 512×512 pixels. This means that the larger the FOV is, the lower the spatial resolution of the resulting images. Thus, one should always try to keep the FOV as small as possible. It should be noted that the segmentation accuracy is always comparable to the pixel size (Testi *et al.*, 2001), so the bigger the pixel the lower the accuracy.

The classification of bones within CT images is traditionally considered an easy task in image processing. Indeed, if one has to segment the outer surface of a cadaver bone scanned while immersed in air or water, it is possible to extract the bone pixels using morphological operators, such as thresholding, erosion, and dilation. However, when the bones are imagined in vivo, the separation between one bone and another might become difficult to define, especially in the elderly, where the inter-capsular space of the joints tends to be smaller, and the presence of osteoarthritis further complicates the picture.

Recently, some complex schemes have been proposed to overcome these difficulties, where region growing with local thresholding is combined with morphological operators (Kang *et al.*, 2003). Another interesting approach uses density–distance augmented Chan–Vese geometric active contours based on the theory of curve evolution and the level-set framework (Truc *et al.*, 2009). However, no extensive studies are available on how robust these algorithms are over large populations. Nowadays the most popular commercial software packages used for bone segmentation (such as Mimics, Materialise NV, Belgium; ScanIP, Simpleware Ltd., UK; Amira, Visage Imaging GmbH, Germany) still provide several segmentation tools including threshold-based, region-growing, morphological, and level-set methods, which suggest that in reality we are still far from full automation.

Another approach to segmentation of bones in CT images involves the use of morphing and atlas-based methods. Elastic registration techniques make it possible to "morph" an existing polygonal surface (surface model) into another shape by minimizing a certain spatial cost function. The simplest case is when one surface is morphed onto another, and the cost function is defined as the distance between the two surfaces. While useful in many cases, this approach does not solve our problem, since the bone geometry is what we are looking for. More relevant here are methods that define the cost function directly onto the CT images. There are various approaches, but in general the 3D scalar field represented by the CT data is transformed into some sort of vector field that drives the shape of the surface model toward a local minimum. This transformation is chosen so that the minimum corresponds to the separation between the bone and the outer soft tissue, so that at convergence the surface model actually segments the bone region in the CT 3D image.

In theory, these methods could be initialized with any initial surface, e.g., a sphere. However, we want the elements of the polygonal surface to avoid excessive distortion; to make the process more efficient, it is often more convenient to start the morphing process from the polygonal surface of another bone of the same type. In this case, the algorithms first make an affine registration, changing the rotation–translation–scaling factors to minimize the distance between the surface model and the underlying 3D image. Then, the surface model is morphed until the cost function goes below a given tolerance.

There is currently a great deal of research in this area, and some recent works claim very good results. Still, it is not so clear whether the results these methods yield are adequate for modeling applications. The whole idea of morphing is that where the information content of the source 3D image is not sufficient to drive the segmentation process automatically, the a-priori information that the template contains can fill the gap. The problem is that this approximation, while generally accurate on average, can be quite inaccurate locally. If the area that is poorly approximated happens to be a region where high stress or strain gradients are induced by loading, this approximation of the local geometry could make a significant difference.

As mentioned, in the rest of this chapter we shall assume that the primary source of information we use to generate our bone model is a CT scan.

4.3 Finite-element mesh generation

4.3.1 Rationale

The automatic generation of finite-element meshes has been a central topic of research in the 1980s and 1990s; that period saw a proliferation of methods in the specialized literature, most of which disappeared soon after having been proposed. For an extensive review of the most important methods for biomedical applications see (Viceconti and Taddei, 2003). Today the problem may be considered solved if (a) domain geometry is represented by high-order mathematical surfaces, and (b) the type of simulation does not impose particular requirements on the type of element topology, and tetrahedral elements can be used. In most whole-bone biomechanics simulations, the second condition is met, while the first in general is not: all the segmentation methods outlined in the previous section generate stacks of planar polylines (chains of rectilinear segments) or polygonal (triangular) surfaces. So it appears reasonable to discuss the various methods used for the automatic mesh generation of bone models from CT data using a taxonomy based on the type of information used to initialize the mesh generation, i.e. to define the domain boundary.

4.3.2 High-order mathematical surfaces

Finite-element modeling software developed mostly in the field of so-called *computer-aided engineering* (CAE), where machine parts had to be tested for strength or other engineering requirements before fabrication. Thus, it became quite natural to assume that for every part to be modeled a *computer-aided design* (CAD) model of its geometry would have been available. Computer-aided design software describes three-dimensional objects using essentially two methods: *constructive solid geometry* (CSG) and *boundary representation* (B-Rep). With CSG, solid bodies are modeled as combinations of Boolean operations on parametric elementary solids, such as spheres, cylinders, or cubes. With B-Rep, a solid is defined as the space limited by its bounding surfaces, which are defined with high-order mathematical representations, such as Bézier, Coons, or non-uniform rational B-Spline (NURBS) patches (Yamaguchi, 2002). It is intuitively clear that CSG works better for regular solids, so biological shapes are almost always modeled using the B-Rep approach. So the problem to be faced in order to capitalize on the wealth of commercial software available is that of converting the polygonal surfaces obtained with the segmentation into some high-order mathematical surfaces.

For many years this problem was hard to solve, and this forced researchers to use crude methods, such as interpolating the stacked contours obtained by segmenting the individual CT slices with a single NURBS patch (Merz *et al.*, 1991). This created problems in the regions where the bone topology bifurcated, as at the epiphyses of the femur.

The problem was finally and rigorously solved by developing methods that would approximate groups of polygonal facets using a NURBS patch. These algorithms were also made available in commercial software, such as Geomagic (Raindrop Inc., USA). While in principle these methods could be fully automatic, to have well-conditioned B-Rep solids they require the user to divide the polygonal surface into portions manually, with each surface fitted with a single NURBS patch.

However, once the B-Rep solid is created, it can easily be imported into the pre-processing module of one of the many finite-element analysis software packages commercially available, and then automatically meshed. In principle, the decomposition into tetrahedral cells of a B-Rep solid is a fully automated process. Still some additional effort might be required to obtain high-quality meshes.

Firstly, if more than one automatic mesh generation (AMG) algorithm is available, it is important to understand the pros and cons of each approach. While AMGs based on the Delaunay triangulation (Viceconti and Taddei, 2003) are still available in many software packages, experience suggests that for meshing bones advancing front AMGs (Viceconti and Taddei, 2003) should be preferred. Advancing front meshers create good quality meshes at the surface, where the gradients are usually highest. The few ill-shaped elements tend to appear in the closure of the advancing front, which is at the core of the body being meshed; since in many bones the core is a medullary cavity, this produces very good results. In addition, it is quite unlikely that this algorithm will generate null-volume tetrahedrons (called *slivers*), whereas with Delaunay AMGs this is quite common.

Another point to consider is the spatial grading of the mesh. The outer surface is the most delicate area, not only because toward the surface the strain gradients tend to be the highest, but also because partial volume effects will inevitably affect the material's remapping (see following sections) on the most superficial elements, and the smaller these elements are, the smaller will be the error involved. Thus, it is recommended that an appropriate spatial grading is imposed on the mesh generator, so that the tetrahedrons near the surface are smaller (Figure 4.3).

A last point is the choice of the element's order. In most commercial solvers, at least two types of tetrahedral cells are available: four-node tetrahedrons (4Tet) and ten-node tetrahedrons (10Tet). A 4Tet element has linear displacements but constant deformation, whereas a 10Tet element has parabolic displacements and linear deformation. Given the large number of elements that modern computers

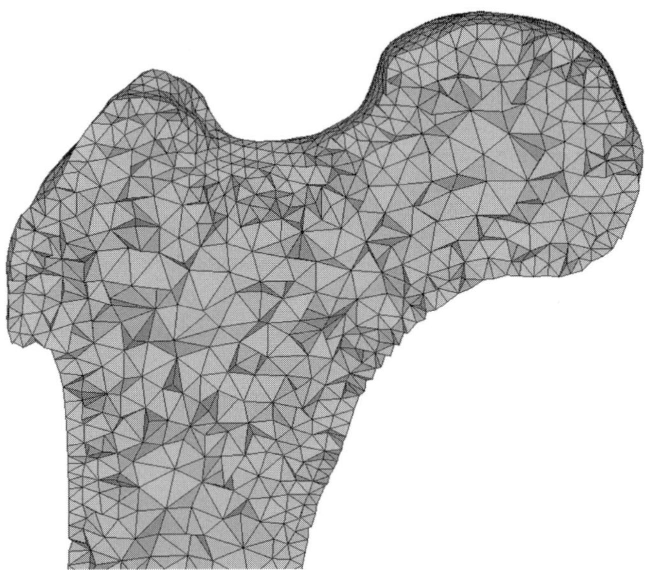

Figure 4.3   A section along a frontal plane of a finite-element mesh of a human femur: the element dimensions are smaller near the surface of the bone and gradually increase toward the internal region.

can handle, one can consider using 4Tet, in sufficient number. However, while for the geometric approximation a linear tetrahedron of a few millimeters size is sufficient to model the geometry of a long bone such as femur or tibia, for strain approximation 10Tet is recommended whenever possible. This implies that one can even generate a 4Tet mesh, and then convert it into a 10Tet mesh by simply adding mid-nodes to the elements' edges. In this way, the mesh could be generated with simpler elements; the higher order would not improve the geometric approximation, but it would significantly improve the strain approximation. Of course all this is presuming that the specific type of simulation does not impose constraints on the element type: for example, in some codes for impact analysis only 4Tet elements are supported.

   A big advantage of having the bone geometry defined as a high-order mathematical entity is that complex operations such as re-meshing come at almost no cost for the user. Thus, in these cases, it is usually quite simple to run a full convergence test over the simulation, to ensure that the level of mesh refinement is adequate for the stress and strain gradients; it is also possible to make adaptive refinements, adding elements where the gradients are higher, but leaving a relatively coarse mesh elsewhere.

4.3.3   Meshing polygonal surfaces

For many years impossible with most commercial codes, the automatic direct generation of tetrahedral meshes from a polygonal surface is now finally becoming possible, without first refitting the mesh with high-order mathematical surfaces. Here there are two possible approaches. The first uses the triangular facets that compose the polygonal surface directly to initialize the advancing front mesher. This approach is potentially dangerous and should be used with great caution. The problem is that most segmentation methods pay no attention to the shape of the triangles being generated, and thus it is quite common that in a polygonal surface many triangles are very small, or very big, or strongly acicular (with very high aspect ratios). It is also possible to have degenerate triangles, although these are relatively easy to detect and remove. This means that before the polygonal surface can be meshed directly, the surface must be re-processed to ensure that the triangles are all well-conditioned and well-shaped.

A second approach uses the polygonal surface simply as a boundary definition, but regenerates both the surface and the solid meshes from scratch. In this case, poor conditioning on a polygonal mesh is of no concern. Another problem, which this second approach also cannot deal with, is that many polygonal surfaces produced by segmentation algorithms are topologically incorrect: in too many cases the surface has holes, duplicate triangles, and even manifold topologies (i.e. when more than two triangles share the same edge). Here the only solution is to use special grid-projection algorithms, such as described in (Taghavi, 1996).

4.3.4   Meshing a set of planar curves

Primarily for historical reasons, I should mention that some methods were proposed in the literature where the mesh, usually made of hexahedral elements, was generated from the stack of planar contours obtained from 2D segmentation, using such mesh generation methods as *mapping* or *sweeping* (Viceconti and Taddei, 2003). This approach was cumbersome and labor-intensive. The curves had to be spaced in relation to the mesh refinement in the direction of the tomographic axis. To solve for bifurcations (e.g. in the proximal femur, where the diaphysis separates into the femoral neck and the greater trochanter), the curves had to be edited manually. Nowadays the method remains in use only for extremely simple, tubular-like, anatomical regions.

4.3.5   Direct meshing of the CT lattice

If we consider the stack of CT images as a 3D image, we can represent it as a lattice, a grid of rectilinear and equally spaced hexahedral elements, where each cell represents an image voxel and carries one Hounsfield unit of attenuation. In principle, we could see this as a hexahedral finite-element mesh, which covers the entire imaging volume. It is trivial to write a program that makes this direct image-to-mesh conversion. However, the problem is that a full resolution CT dataset of a long bone can have 250 slices, each with 512×512 pixels, which makes 65 million elements, in comparison to the 100 000 elements that typically form a bone finite-element model generated using the previously described methods. Until recently, problems of this large computational size were not treatable unless with very special solvers that took advantage of the fact that all elements are identical in shape and size, and that ran on very expensive supercomputers (Van Rietbergen *et al.*, 1996). Alternatively, 3D images were aggressively sub-sampled, to make the number of voxels or elements computationally tractable with the solvers and the computers commonly available (Keyak *et al.*, 1990). Unfortunately, this latter approach drastically reduces the predictive accuracy of these models (Viceconti *et al.*, 1998).

However, the scenario is changing. Multi-core computers equipped with very large amounts of memory are now common and intense research has produced new methods for solving very large sparse systems of linear equations that scale well when parallelized on a large number of processors. As a result, direct meshing, sometimes called image-based meshing or voxel meshing, is also becoming a viable option for organ-level simulations. Another possibility is the use of meshless methods that are initialized directly with the 3D image. An example of such a method has been provided in Chapter 2, where I described a numerical method alternative to the finite-element method, called the meshless-cells method (MCM).

4.4   Modeling the boundary conditions

4.4.1   Organ-level boundary conditions

Each bone is subjected to a complex load system transmitted by adjacent bones, muscle tendons, ligaments, and other connective tissues, such as fasciae. Bones are also loaded by transverse actions produced by the expansion of muscles during their contraction. In some cases, bones are also exposed to loads external to the body, such as the impact load produced by a fall.

To model these boundary conditions, we need to define how each of these actions is applied to the bone, and its direction and intensity over time, and the eventual presence of additional boundary conditions, such as constraints, to ensure that the model is in static equilibrium at each instant.

4.4.2   Application models

The forces transmitted by adjacent bones usually pass through a joint. Under physiological conditions the joint cartilages ensure that the joint load is spread over a sufficiently large portion of the articular surface, which avoids localized overstress of the cartilage itself and of the bone underneath (the subchondral bone). Thus, the articular force is usually applied to the bone as a distribution of pressure over a fairly large area. If the model we are building has to predict the stress and strains near the articular surface, i.e. in the subchondral bone, then it is important to ensure that this pressure distribution is the most realistic possible, and specific measurements should be made. Alternatively, one can consider defining the load with a coupled-joint intact model, where the pressure distribution acting on the bone is the result of a contact simulation. However, if we assume Saint-Venant's principle to be valid, i.e. if we are interested in the stress and strains of the bone only in regions that are distant from the joint surface by at least one or two times the size of the contact area, then we can replace the pressure distribution with its resultant. Since the friction in natural joints is very small, we can always assume that the joint reaction passes through the instantaneous center of rotation of the articulation. Thus, assuming that Saint-Venant's principle is valid, the force transmitted through an articulation can be represented as a single force concentrated at the joint center, or at the point where the force vector pointing toward the joint center intersects the articular surface.

Similar reasoning should also be applied for muscle forces. Muscles tendons insert into bones over fairly large areas, in some cases very large ones. In principle, we should distribute the muscle force over such an area, with a spread function that is, however, nearly impossible to determine. One could assume that the force is spread evenly over the entire insertion surface. Specific studies have shown that such uniform spread produces strains that only differ significantly from those produced by a simpler concentrated force applied at the centroid of the insertion area near the insertions (Polgar *et al.*, 2003). But if there is no specific need to model the stress and strain fields accurately near the muscle insertion, there is no evident advantage in distributing the muscle force over the insertion area using an arbitrary function. The increasing automation at the whole-body level might help in creating multiscale models, where dozens of independent actuators are modeled for each muscle; this would produce more anatomical boundary conditions than the single force per muscle bundle that we use nowadays; in this case we would have a non-constant distribution of forces over the muscle insertion area, which would be worth accounting for.

Passive structures (ligaments, capsule, and fasciae) are usually modeled as constraints rather than as active boundary conditions. Their application is usually limited to small portions of the bone surface, and thus they are frequently modeled with concentrated boundary conditions.

4.4.3 Intensity and direction over time

There are four possible approaches to defining the intensity and direction of the forces acting on a bone:

- Define them by completely ignoring the musculoskeletal physiology, imposing oversimplified boundary conditions that produce in the bone stress and strain fields of form and intensity roughly comparable to that found on physiological conditions. This approach is used to investigate inherent properties of a series of bones, e.g. to compute the average flexural rigidity of a long bone, such as the tibia. This case also includes the models used to investigate the ability of the bone to stand external loads, which are usually modeled as simple loading conditions applied directly to the bone of interest, and assuming that the rest of the forces (articular, muscular, etc.) can be neglected.

- Define them by imposing an aggressive simplification of the loading conditions that *on average* are acting on that bone during a certain standardized motor task. The task of interest is analyzed at the body level over a large number of subjects, and a typical pattern of muscle and joint forces is derived. A simplified load set is then defined by finding a reduced set of forces that are statistically equivalent (or sufficiently close) to this average force pattern. No more refined approach would make sense, since the approximation we make in transposing the muscle forces during a given task as predicted in one subject or averaged over a group of subjects onto the bone of another subject is so crude that trying to be detailed elsewhere in the model would make no sense.
- Define intensity and direction of loads as statistical variables. The anatomical conformation of the joints tends to limit the range of possible directions this resultant can take: in the hip, experimental measurements in vivo confirmed that the hip joint reaction always remains within a cone of 20–30 degrees, during a disparate list of motor tasks. Thus, one can replace the hip joint reaction with a force resultant acting on the femoral head center, and whose intensity and direction are defined as randomized variables with pre-defined limits, i.e., a truncated Gaussian distribution. Of course the model predictions would be statistical variables as well, but if the coefficient of variation is not too large, such information could be useful in a number of cases.
- Create a coupled multiscale model where the musculoarticular forces are computed at the body level, as explained in Chapter 3, and then applied to the organ-level model of the bone described in this chapter. This is the most challenging approach, but also the one that promises the greater accuracy in prediction.

4.4.4 Additional boundary conditions

Given that the system of forces acting on a bone at each instant is in instantaneous equilibrium, if we could reproduce such a load set in our organ model, there would be no rigid body motion left. However, depending on the level of idealization that is introduced when modeling the external forces, we might end up with a load set that is not in static equilibrium. In these cases it is necessary to add a constraint, usually by imposing zero displacement to some nodes on the surface of the bone model. Great care should be taken over such an operation. The constraint result force should be small, possibly significantly smaller than the external forces acting on the bone. In addition, the location of the constraints should ensure that the bending and torsional moments that such external forces impose on the bone are not too different. Last, but not least, the constraints should be placed far from the region of interest, as the boundary effects might significantly alter the stress and strain fields in the region near the constraints.

Another possibility is to impose compliant constraints so that the bone would move rigidly to a spatial configuration that is in equilibrium with all forces applied (including the elastic force of the constraints). While this approach might be useful in some cases, we should never forget that this does not dispense us from the responsibility of choosing constraints wisely in relation to the idealization we make of the boundary conditions. When compliant constraints are used, bones should have only very small displacements, and the constraints' reaction forces should remain small.

4.5   Constitutive equations of bone tissue at the continuum level

4.5.1   Generalities

The problem of defining a constitutive equation for bone at the continuum level has been and still is a major research topic in the specialized literature. The debate is partly motivated by an objective complexity of the argument, but it also certainly arises from an epistemological misunderstanding between practitioners.

In the field, we see two clashing attitudes: the first group sees any idealization as an evil, and values constitutive models for the least content of idealization they have. In other words, the more complex a model is and the better it accounts for as many observed phenomena as possible, the better it is. This approach springs from the area of applied mathematics, whose primary goal is the increase of knowledge, in the broadest sense. The second approach instead is pragmatic and derives from the area of engineering, whose scope is problem solving. Here the goal is to find the simplest constitutive model capable of allowing predictions of the phenomena of interest with the expected level of accuracy. It is clear that while both positions are useful and legitimate, and in a way complement each other, they tend to be quite incompatible in terms of scope. I want here to take my stand clearly in the second camp: the scope of this chapter, and of this book in general, is to work on the solution of practical problems of bioengineering, and not on the increase of knowledge for the sake of it. The theories I shall present here do not necessarily have the gift of intellectual elegance and indeed do not aim to be exhaustive; as soon as we see that a particular phenomenon produces negligible effects on the problem of interest we shall be quick to ignore it through appropriate idealization.

Our primary scope is to assess the biomechanical competence of bones, i.e. to evaluate the intensity of a given loading case that produces failure. This involves three different scenarios, each imposing different constraints on the constitutive equation we use to describe the bone material at the continuum level. First of all, we need to predict accurately the elastic behavior of the bone, e.g. how the bone deforms under the action of stresses that are well below the elastic limit. Secondly, we need to model those fractures where tensile stresses play an important role. Some examples are the low-energy fracture of the femoral neck in osteoporotic patients, or the spiral fracture of the tibia in younger subjects. Since these fractures are predominantly fragile, in these cases the bone does not exhibit any post-elastic behavior, and the constitutive equation can still be limited to the elastic range, although it has to be associated with a failure criterion, used to decide when the elastic limit is reached. Lastly, we need to model those fractures where the predominant stress is compressive, and the bone fails by being progressively crushed under the action of the load. In this case we need a much more complex constitutive equation, which also accounts for this post-elastic phase.

4.5.2   A digression on terminology

We all use in day-to-day conversation such sentences as, "Bone is a viscoelastic material." I would strongly recommend that we stop using these expressions, as they imply a totally erroneous approach. A material is not viscoelastic, orthotropic, homogeneous; a material is what it is. What we really mean by saying, "Bone is a viscoelastic material" is that if I use a viscoelastic constitutive equation I can predict the biomechanical behavior of bone tissue with an accuracy that is adequate to my purpose.

The reason I am making this point is that too often when one presents a simplified constitutive equation some colleagues criticize it on the basis that, "You did not take into account X, but bone is an X material." It seems almost as though a model that neglects X is an idealization, whereas a model that includes X is "true." This objection does not make any sense; it should rather be replaced by a much more complex set of questions:

- What is the scope of your model?
- What is the accuracy that such scope requires for your model's predictions?
- Under the idealization you propose, what is the accuracy you can achieve?

The reality is that every model is an idealization of physical reality, which by definition is infinitely complex, and thus can only be investigated through some idealizations, which reduce the complexity to a finite level. We can use more or less aggressive idealizations, depending on the scope of the model and on the predictive accuracy we need.

So please do not say, "Bone is a viscoelastic material;" say, "I model bone as a viscoelastic material, because this ensures that the predictive accuracy of my model is sufficient for my modeling scope." It is a bit longer, but much more accurate.

4.5.3   The elastic constitutive equation

The constitutive equation we are looking for should provide accurate predictions in the elastic range, under small deformations and small displacements (actually this last requirement can be removed later on by using iterative schemes to account for large displacements while still using linear constitutive equations). Such infinitesimal linear elastic constitutive equations will make it possible to predict displacements, stresses, and strains of whole bones subjected to loads small enough that the tissue remains well below the elastic limit.

As stated in Section 2.3.4, Hooke first proposed the original idea of elastic behavior, where stresses are linearly dependent on strains, but it was generalized to tensors by Cauchy in the following form:

$$\sigma_{ij} = D_{ijkl}\, e_{kl}, \tag{4.1}$$

where σ_{ij} is the stress tensor, e_{kl} is the infinitesimal strain tensor, obtained under the assumption of small deformations and small displacements, and D_{ijkl} is called the elastic modulus tensor, and has 81 elements, only 36 of which are independent, thanks to symmetries on the ij and kl indices. If we can postulate the existence of a strain energy density function

$$W = \frac{1}{2} D_{ijkl}\, e_{ij} e_{kl}, \tag{4.2}$$

so that

$$\frac{\partial W}{\partial e_{ij}} = \sigma_{ij}, \tag{4.3}$$

which is the case most of the time, then we can assume that the quadratic form that defines the strain energy density function is also symmetric, and thus $D_{ijkl} = D_{klij}$, in which case the number of independent constants goes down to 21. Thus, in the most general case we need to identify experimentally 21 independent material constants. Fortunately, most materials, including bone, show additional symmetries, which significantly reduces the number of these constants. In most cases the elastic behavior of the material is symmetric over three mutually orthogonal planes, and thus brings down the number of constants to nine. If we assume the material to be isotropic (i.e. that its elastic behavior is identical in every spatial direction), then the number of independent elastic constants is reduced to two.

A second dimension of the problem is that in most engineering materials, the composition and the microstructure of the material are sufficiently homogeneous that we may assume that such elastic constants are the same everywhere; in biological materials, however, this assumption is rarely acceptable. In bone in particular, the elasticity of the tissue is strongly dependent on the degree of mineralization, which can vary from point to point and over time.

A third problem we need to address is that bone tissue has a viscous behavior that in some cases is non-negligible, which means that the stress is not only a function of the strain, but also of the strain rate. If we describe a body in Cartesian coordinates, and assume small deformations and small displacements, we can define as *linearly viscoelastic* a material whose behavior is well represented by the following equation:

$$\sigma_{ij}(x,t) = \int_{-\infty}^{t} G_{ijkl}(x,t-\tau)\frac{\partial e_{kl}}{\partial \tau}(x,\tau)\,\mathrm{d}\tau. \tag{4.4}$$

Thus, the questions we are called to answer are:

- Can we assume that the viscous effects are negligible, and if not how can we account for them?

- Can we assume that bone is a homogeneous material, and if not how can we describe the constitutive equation as a function of space–time?
- Can we assume that bone is an isotropic material, and if not how can we account for its anisotropy?

4.5.3.1 Viscoelasticity

When modeling bone biomechanics at the organ level under physiological or pathological loading conditions that can be observed in vivo, there are four possible scenarios:

(a) The load rate is so low that viscoelastic phenomena can be neglected.
(b) The load rate is in an intermediate range where viscoelastic phenomena cannot be neglected, but we can account for them simply by adding a correction factor that links the modulus of elasticity to an average strain rate, while retaining a full quasi-static formulation of elasticity.
(c) The load rate is so high that the whole loading event must be modeled as a transient dynamics process, where bone is assumed to be fully viscoelastic.
(d) The load rate is null, the loads are kept constant for large time intervals, during which there are non-negligible processes of creep (if stress is kept constant) or relaxation (if strain is kept constant).

In case (b) we model the bone loading as a quasi-static process; the empirical equation used to express the relationship between the tissue density and its modulus of elasticity (see the following section) is extended with a term that accounts for the average strain rate (Carter and Hayes, 1977):

$$E = 68\,\dot{\varepsilon}^{0.06}\,\rho^{2}. \tag{4.5}$$

Case (c) requires a completely different modeling approach, not only because of the viscoleasticity, but also because of the transient dynamics; these problems are usually quite difficult to solve using classical implicit solvers, and are typically approached using codes specifically developed for impact analysis, most of which rely on explicit solvers, a particular type of finite-element analysis solver developed specifically to solve impact problems.

Case (d) is of interest only in a limited number of cases. While bone creep and relaxation phenomena might be relevant in some very specific problems (such as distraction osteogenesis, or press-fit modeling in cementless joint replacements), when the aim is to model the whole musculoskeletal system one should consider that the viscoelastic behavior of cartilage, ligaments, and tendons would surely produce much larger effects than those of bone, which can thus in many cases be neglected.

4.5.3.2 Tissue inhomogeneity

In human bones there are two types of tissue inhomogeneity. The first is that due to the presence of various histological types of tissues. In all bones there are lamellar compact and cancellous bone tissues; in long bones there is also osteonal compact tissue. As discussed previously, histological types cannot be directly discriminated from CT data. While compact bone is usually denser than cancellous bone, for various reasons (including partial volume effects) the density distributions of the various tissue types overlap. In addition, there is no way of distinguishing between lamellar and osteonal compact bone on a purely densitometric basis. Thus the regions of the various histological types can be separated in the CT images only by adding some a-priori knowledge on the anatomical location of the various tissue types, and this is done during segmentation.

There are essentially two approaches. The more complicated one separates compact lamellar, compact osteonal, and cancellous bone types into regions, each modeled differently. The compact lamellar region, which is usually a thin shell around the cancellous bone region, is modeled explicitly using plate elements (Imai *et al.*, 2006), which exhibit both flexural and membranal stiffness. In this case one must be careful to ensure that the shell elements and the underlying solid elements used to model cancellous bone have compatible displacement modes. The modulus of elasticity and the thickness of the plate elements are usually assumed constant, and are defined on the basis of average values from population studies. Compact osteonal and cancellous bone types are modeled with solid elements (usually tetrahedrons) as an elastic material with density-dependent elasticity modulus. They are treated separately because distinct density–elasticity equations are used for the two histological tissue types, but also because the relatively simple geometry of diaphyses in long bones (where all compact osteonal bone is found) makes it easy to define, a priori, an orthotropic constitutive equation with the material's highest stiffness direction oriented along the diaphysial axis (Yosibash *et al.*, 2010).

The second approach refuses to discriminate between histological types on the basis of densitometric information, and models bone as a unique solid continuum, locally isotropic and inhomogeneous, whose modulus of elasticity is defined in each point by the local value of mineral density and then averaged over the volume on an element-by-element basis (Taddei *et al.*, 2007). An open-source implementation of material mapping code is available (Bonemat©, SCS, Italy) (Figure 4.4).

It is also possible to implement a more sophisticated scheme, where the modulus of elasticity varies across each element: this can be done by using modified element formulations (Edidin, 1991), or more simply by using a pseudo-temperature field to model the spatial gradient of elastic properties (Helgason *et al.*, 2008). While these methods provide significant advantages when large elements are used, thanks to the computational power of modern computers in most cases very small elements (having side lengths of the order of 2 mm) are used, and in such case it is not clear if the additional complication these methods involve is worth the gain in accuracy when compared with the simpler element-by-element averaging scheme.

When using this approach, particular attention is only given to ensure that near the outer surface the mesh is highly refined, so as to capture, elemen-by-element, the gradients of tissue density as we pass from the external shell of compact lamellar bone to the cancellous bone interior. In validation studies on the femur, this much simpler approach has yielded results that are equal to or better than those

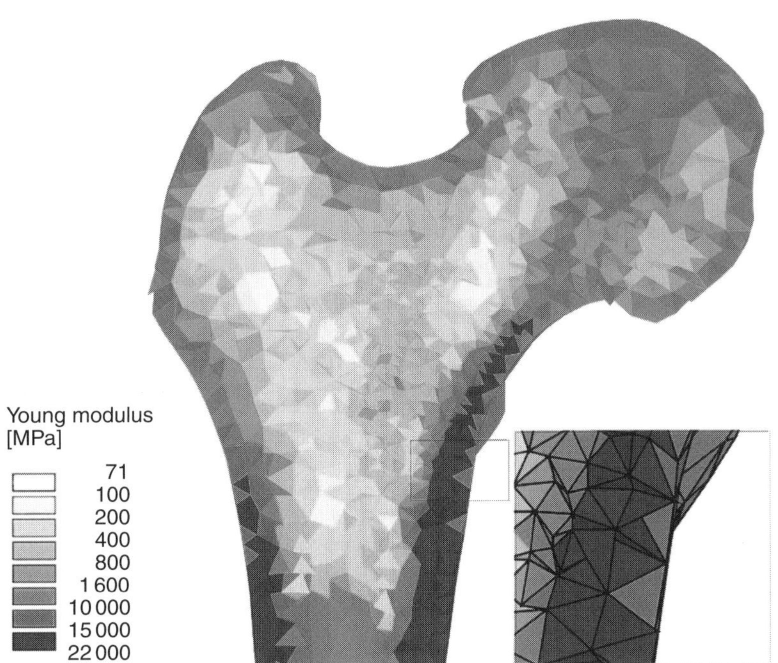

Figure 4.4 Inhomogeneous material properties mapped onto a finite-element model of a proximal femur using the Bonemat© Software.

obtained with the previous much more complicated approach (Schileo *et al.*, 2007), and thus it should be preferred, in my opinion. For the vertebra, the local biomechanics is probably different and the role of the outer shell of plate elements might be non-negligible; recent validation studies which include the plate shell predicted experimental measurements very accurately (Imai *et al.*, 2006; Matsumoto *et al.*, 2009).

4.5.3.3 Tissue anisotropy

Speaking of tissue anisotropy in general is meaningless; we need to specify at which dimensional scale we are looking. The use of inhomogeneous constitutive equations ensures long-range anisotropy. So eventually the problem is to account for local, short-range anisotropy, which must be modeled with the single element. The problem is not in assigning anisotropic elastic properties, something almost every finite-element code can easily do nowadays, but in defining, for each element, the degree of anisotropy, and the orientation of the principal axes of the elasticity tensor. In principle one could chop a bone into many small pieces, and measure the orientation and the degree of anisotropy of each piece, so as to create a 3D map of tissue anisotropy (various authors have actually done this; (Brown and Ferguson, 1980)). But this approach would be destructive, as well as overwhelmingly complex and tedious. When non-destructive methods are required, and the source of the information is the CT dataset (which provides a scalar field) it is not self-evident how to find such information on tissue anisotropy. Some exploratory work is being done on the use of density gradients as a source of information (Tabor and Rokita, 2007).

It remains to be seen whether the complication is worth it. In general, studies aimed at predicting the global strain field of the whole bone, changing from isotropic to orthotropic, produce small changes in predicted strains (Baca *et al.*, 2008; Yosibash *et al.*, 2010). However, anisotropy might play a major role in locally driven processes, such as fracture initiation.

4.5.3.4   Meshless methods

It is evident that all these subtle methodological differences would lose any significance in a scenario using a meshless scheme, like the MCM described in a previous chapter, or when the computation power is sufficient to push the resolution of these methods to the CT voxel. In such cases all the information available from the CT scan would be entirely captured by the model; the anisotropy that emerges from the inhomogeneity would have a range of a millimeter or less, depending on the CT spatial resolution (and trying push any further than this would be unwise, since for bone the assumption of a continuum starts to fail grossly around the 100–300 micron mark).

4.5.4   Post-elastic constitutive equations

As mentioned, if the bone fracture is dominated by tensile stresses, the fracture is usually fragile, and no appreciable post-elastic deformation is observed. The bone deforms elastically until the failure criterion is met, and then fails without absorbing any additional energy.

In other cases, however, the bone fails under predominantly compressive loads. Typical examples are the compression fractures of vertebral bodies. Here the force–displacement curve measured during compression tests on cadaver vertebrae shows a linear tract, a clear post-elastic phase where significant energy is absorbed, and a re-stiffening phase, when the cancellous bone has been crushed enough to fill most porosities and the tissue starts to behave like a compact continuum. Another case where a purely elastic constitutive equation is likely to produce excessive errors is the simulation of so-called *greenstick fractures*, which are seen mostly in children.

The definition of a constitutive equation that accurately predicts this entire post-elastic behavior is beyond the scope of this book. While the understanding of bone behavior under large plastic strains is surely of general interest, if the aim of the model is to decide if and when a bone *starts* to fail, a much more limited constitutive equation is required. What we need is a post-elastic constitutive equation that accounts for the fact that because of large inhomogeneities in bone density, especially in pathological bones, during compression we might have small portions of bone that are well above yield, whereas the vast majority of the tissue is still acting under a purely elastic regime. In this perspective, the constitutive equation should be able to model the post-elastic deformation of these low-density regions, so as to avoid the totally unrealistic stresses that any model would predict if these high deformation volumes were assumed to remain indefinitely elastic.

4.6 Verification and validation of whole-bone models

4.6.1 Introduction

If we plan to use organ models to make predictions of any usefulness we should know what level of accuracy we can expect from our model. In Chapter 1, I discussed the general problem of assessing modeling technology to be used in clinical practice. Here, I shall specify the details that are relevant for models at the organ scale.

4.6.2 Model selection and verification

Model selection remains the most critical step, and also the one that is most difficult to formulate into a general good practice. In this chapter, we assumed that the modeling objective is to predict the stress, strain, the displacement of a bone loaded below its elastic limit, or the load required to bring such bone to fracture, given the direction and orientation of the loading conditions. For these objectives, a linear elastic model of the bone modeled as a non-homogeneous continuum is usually adequate. Eventually, one might improve the predictive accuracy by including large displacement non-linearity when long bones are subjected to bending loads where small strains can still induce large displacements, or by modeling the non-linear post-elastic behavior of the bone tissue immediately after yield.

But a bone model can be used to answer many other questions, and for each of them the modeling assumptions should be revised with care. For example, if one wants to model the biomechanical interaction between the bone and an implanted device, such as a joint prosthesis, it is usually necessary to account for the non-linear large sliding frictional contact that develops between the implant and the host bone. If one is modeling the spine, it is usually necessary to account, explicitly or implicitly, for the strongly non-linear behavior of the intervertebral disk. If this is done explicitly the model should probably include a poro-elastic model for the complex interaction between the disk matrix and the fluid that saturates it.

If one is uncertain whether a given idealization is going to produce large differences in the prediction, the best way to verify it is to compare the results of two models, one with and the other without the idealization. Alas, this is a quite expensive approach and to be effective it should be done in the context of a validation experiment that can provide some base truth, against which we compare the effect of different idealizations. This approach is part of the general vision of scientific inquiry called *strong inference*, discussed in Chapter 1. In this vision one should never face a scientific inquiry with only one hypothesis, but should rather start with a battery of possible hypotheses, that should then be subjected to extensive experimental validation in order to be falsified. The hypothesis that survives this process (if any) is the best candidate, in the classical abduction process (see Chapter 1).

Of course, in many cases the problem of interest has already been extensively investigated in the previous literature, and thus one can adopt a particular idealization of the problem according to some pre-defined best practices. However, caution should be taken in adopting methods used in previous studies, if these methods have not been thoroughly tested. The fact that someone else has used a method in a published study does not automatically guarantee that the method is sufficiently accurate for that particular modeling objective. Never forget, as scientists we do not work to be right, we work to find the truth, even if this means disproving our beloved theory or method. Never fall in love with your own ideas!

In the context of finite-element analysis, verification (i.e. assuring that the mathematical model is solved numerically with sufficient accuracy) is either very simple or very complex. Verification involves both the code and the model. If you use a commercial code its developers should have already verified the finite-element software code. Check for the code benchmarks, and if you have any doubt resort to vendor-independent benchmarks such as those provided by NAFEMS.[1] Verification of linear finite-element models is possible nowadays with moderate effort. Best practice, which is trivial if one uses automatic mesh generation, is to conduct a mesh refinement. The domain is initially meshed with a fairly coarse mesh, and then solved for the result variable that is of greater interest in achieving the modeling objective. The operation is repeated many times, progressively reducing the average element size of the mesh. Under a number of conditions that are usually satisfied in linear elastic problems like the ones dealt with in this chapter, as the mesh refinement increases above a certain mesh refinement, the predicted values should converge monotonically and asymptotically to the exact solution.

In principle, mesh convergence could be monitored on the peak value predicted for certain model output; however, for a number of practical considerations, it is more convenient in many cases to monitor the mesh convergence at a finite number of points, strategically located where we expect to have the largest physically relevant strain gradients (by "physically relevant" I mean those gradients that are part of the physical problem and not artifacts of our idealizations; if concentrated forces are present in the model near the node of application, the gradients might be very high, but this is an artifact induced by the idealization of the boundary condition). To do so, one should ensure that each mesh node, or element Gauss point, is constantly located in the same spatial position in all mesh refinements. Alternatively, one could use element interpolation to obtain the result at the same spatial coordinates regardless of the mesh topology, but there is a risk that the interpolation errors might exceed the numerical solution error.

It is usually better to monitor mesh refinement of higher-order results such as strain, as a mesh refinement that appears to be at convergence over displacement may be still far from convergence over strain.

---

[1] www.nafems.org/

When the level of mesh refinement is sufficient to show asymptotic convergence, the change in quantity between two successive mesh refinements can provide a quantitative indicator on which one can decide whether the mesh refinement is sufficient to the modeling objective. Alternatively, one can use post-hoc indicators. These indicators were developed mostly in the context of adaptive mesh refinement, where the program automatically decides in which regions it is necessary to refine the mesh and where this is not required, on the basis of these indicators. There are two types of post-hoc indicator method: flux projection (or superconvergence-based) and residual-type. A detailed discussion of these mathematical methods is beyond the scope of this book, but a recent review can be found by Grätsch and Bathe (2005). Nowadays, most classical flux-projection indicators are available in almost all commercial finite-element modeling software. However, a word of caution should be given, since most of these indicators work under the assumption that the body being modeled is made of a single material; thus, the practical utility for these methods when inhomogeneous continua are involved, as in our case, is limited.

However, when a model is non-linear, the number of options available for verification is quite limited. Mesh convergence testing is still possible, but nothing guarantees that the curve will monotonically converge to the exact solution. In most cases, the only option is to look at the residual of the Newton–Raphson iteration, but the translation of such residuals into a quantified error on the model predictions might be quite complex. In general every type of non-linearity involves a way of estimating, directly or indirectly, the numerical accuracy of the solution, but the problem is discussed only in the most specialized literature.

4.6.3   Sensitivity analysis and inter-subject variability

Given that in most biomedical applications the input parameters of our models can be measured only with significant uncertainty, it is always a good idea to consider a sensitivity analysis. In addition, for non-linear models the sensitivity analysis can also show some singularity in the numerical implementation, which might not show up with a single input set.

The two most common methods of sensitivity analysis of finite-element models use limited methods such as the *design-of-experiments* (DoE), or full-blown Monte-Carlo analyses.

In the simplest DoE implementation (see for example (Chang *et al.*, 2001; Yao *et al.*, 2008)), given by Genichi Taguchi, one explores the hyperspace of the input variables by taking only the minimum and maximum values that the variable can assume. This creates 2^n possible input sets, which correspond to all possible permutations of low and high values. With eight variables, a full DoE requires that the model be run 256 times, each time with different inputs set; for five variable only 32 runs are required. If the uncertainty on the input variable provides a non-truncated normal distribution characterized by mean and standard deviation, the low and high values are fixed by choosing a number of standard deviations. In a normal distribution, 95% of the values lie within two standard deviations, and 99.7% within three standard deviations.

Monte-Carlo methods (see as examples Laz *et al.* (2007); Taddei *et al.* (2007); Viceconti *et al.* (2006a)) are a class of computational algorithms that rely on repeated random sampling to compute their results. In their simplest implementation, each input is defined as a statistical variable characterized by a certain distribution; at each run a random value for each input is generated according to its probability distribution, and the resulting set of inputs is used to run the model. If the number of runs is sufficiently large, the sampling of the inputs and their relative outputs will converge to their probability distributions, thus providing an estimate of the probability distribution for the model's predictions.

Design-of-experiments methods work on very strong assumptions, orthogonal parameter space and linear effects being the most important; assumptions that are acceptable only in the simplest cases. In addition, their computational cost increases exponentially with the number of variables, so for large numbers of variables it is not so convenient anymore. Design-of-experiments methods should be used for relatively simple models, where the inputs tend to have a linear effect on the output, and their number is limited. They can also be used as a first exploratory check to determine whether more computationally expensive methods are necessary.

Monte-Carlo methods make no assumption about the model and its dependence on the inputs, and their computational cost is in general insensitive to the number of variables, although it is always very high: depending on the problem a Monte-Carlo analysis can require hundreds or even thousands of runs. The study of probabilistic methods is quite intense, and some sophisticated methods, such as the *advanced mean value* method can in some cases drastically reduce the computational cost of probabilistic simulations (Laz *et al.* 2007).

4.6.4   Validation against controlled experiments

According to Popper, a theory (and thus a model) can never be conclusively proven true, but can be conclusively proven false (Popper, 1992). Thus, comparing the predictions of our model with the results of controlled experiments does not, strictly speaking, validate the model, and at most fails to falsify it.

This is not to say that validation is a futile exercise; on the contrary, I have repeatedly stated that no model should be used to draw clinically relevant conclusions if it has not been thoroughly validated (Viceconti *et al.*, 2005). However, the reader should be warned that designing and conducting validation experiments is in most cases much more difficult, and much more delicate that developing the model itself. This is a possible explanation of why the scientific literature is full of models, but only a small fraction of them are extensively validated.

The first problem is methodological in nature. Ideally, we should use our model to predict reality, but we have no control on reality. Thus, we resort to a controlled experiment, which is in itself a model of reality. Thus, we use a model to validate another model. This is fine as far as (a) we design the controlled experiment so that it faithfully reproduces those features of reality that are the most relevant for the scope of our model; and (b) we avoid extrapolating the validity of our model beyond the limits of the controlled experiment we used to validate it.

Whole-bone models are typically validated by generating the model from a bone, and then subjecting the same bone to known loading conditions, while measuring various biomechanical quantities. The same quantities are predicted with the model under the same boundary conditions, and compared with experimental measurements.

Whole-bone biomechanical models are usually validated with controlled experiments on cadaver bones. This relies on the assumption that the biomechanical properties of the whole bone do not significantly change ex vivo if the bone has been properly preserved. Fresh-frozen bone, subjected to a thawing cycle in water at 37 °C, for sufficiently long to re-establish proper tissue hydration was found to have elastic properties very close to fresh bone.

The loading set that acts on bone in vivo is extremely complex, owing to the action of joints, musculotendinous units, and other connective tissues, such as fasciae. Accurate reproduction of such loading is very difficult, if not impossible. But in addition, one must always consider that the more complex the experiment is, the less controlled it becomes. In biomechanics experiments, a single force experiment can be controlled with a very high level of accuracy; as the number of independent forces increases, the accuracy with which the total loading is applied quickly degrades. Thus, there is a trade-off between the realism and the control of such an experiment. The use of simple but well-controlled loading conditions is recommended, ensuring they replicate the direction and intensity of the aggregated structural conditions, in terms of axial, flexural, and torsional components. Also, the loading should not be defined separately from the constraints. On the contrary; the loading–constraining system should be designed so that it reproduces the desired axial, flexural, and torsional structural conditions in the region of interest.

During the test, the forces applied externally to the bone are measured using load cells; in addition, the global structural displacement of the bone in the direction of loading is usually measured using linear voltage displacement transducers (LVDT). In addition, in the validation of biomechanical models, displacements of selected points, pressures, and deformations are measured. Of course, all these quantities can be measured only on the outer surface of the bone. While this might seem a limitation, the periosteal surface is almost always the location for the highest strain gradients, i.e. the regions where the model accuracy tends to degrade, so verifying the model accuracy in those critical zones is usually a good idea.

Structural displacements are relatively easy to measure, using contact dial gages, or other types of linear displacement transducer. Displacements are usually measured around where the peak displacement is expected, but also near the regions that in the model have imposed null displacements, so as to verify that such conditions are realistic.

Pressures are relevant validation quantities only when the models account for some contact interaction; they can be measured with pressure-sensitive films, or with pressure transducers that have to be embedded in one of the two contact surfaces, although this might perturb the contact itself.

Strain can be measured using full-field techniques such as reflection photo-elasticity or other optical methods, or using local sensors (strain gages) that measure the surface strain in a very small region. Traditional full-field methods, such as reflective photo-elasticity, are usually complex to use, require delicate preparations of the bone surface that frequently tend to alter the bone tissue elastic properties, and provide a full-field measurement but with moderate accuracy (Cristofolini *et al.*, 1994; Glisson *et al.*, 2000). Strain gages, instead, do not require any special superficial treatment that can alter the bone tissue properties, and provide a very accurate measurement, although in a single location. Although some authors question the reliability of these sensors when applied to cadaver bones, by choosing an appropriate grid size, and accurately following an optimized application protocol, strain gages can work on fresh wet bone with very high accuracy and reliability (Cristofolini, 1997). New optical methods based on digital image processing promise full-field measurements of displacement and strain at accuracies comparable to that of strain gages, but extensive validation studies are still missing.

Of course, to compare the experimental measurements with the model's prediction, we need to define, to a good approximation, the transformation of the reference system between the laboratory coordinates and the model's coordinates. This is called model-to-experiment registration. In whole-bone biomechanics experiments, it was found that the best way is to use an accurate 3D digitizer, mounted on an articulated arm, and collect a large number of points on the bone surface, in addition to the 3D coordinates of all relevant points (center-of-strain gages, location-of-displacement sensors, axis-of-loading actuators, etc.). The measured points are registered to the 3D surface of the bone in the model using an *interactive-closest-point* (ICP) algorithm, which computes the transformation of reference systems that minimize the average distance between the point and the model surface.

4.6.5   An example: validation of a proximal femur model

As an example, I summarize here a series of experiments conducted in our group to validate the protocol for generating patient-specific finite elements of human femurs from CT data (Figure 4.5). Whole femoral bones were obtained through donor programs from donors without any skeletal deformity or skeletal disease. Each femur was scanned over its whole length using clinical computer tomography (CT) with parameters and settings typical of in-vivo imaging; the soft tissue around the bone was simulated by scanning the bones while immersed in water. After the scan, the distal portion of each femur was potted into acrylic cement and encased in a rigid steel box. The femurs were connected to rosette strain gages at various locations on the proximal and diaphysial surfaces. Linear voltage displacement transducers were located near the distal constraint and in the proximal region, where maximum deflection was expected. These instrumented specimens were loaded well below the elastic limit in multiple directions, each simulating a typical loading condition

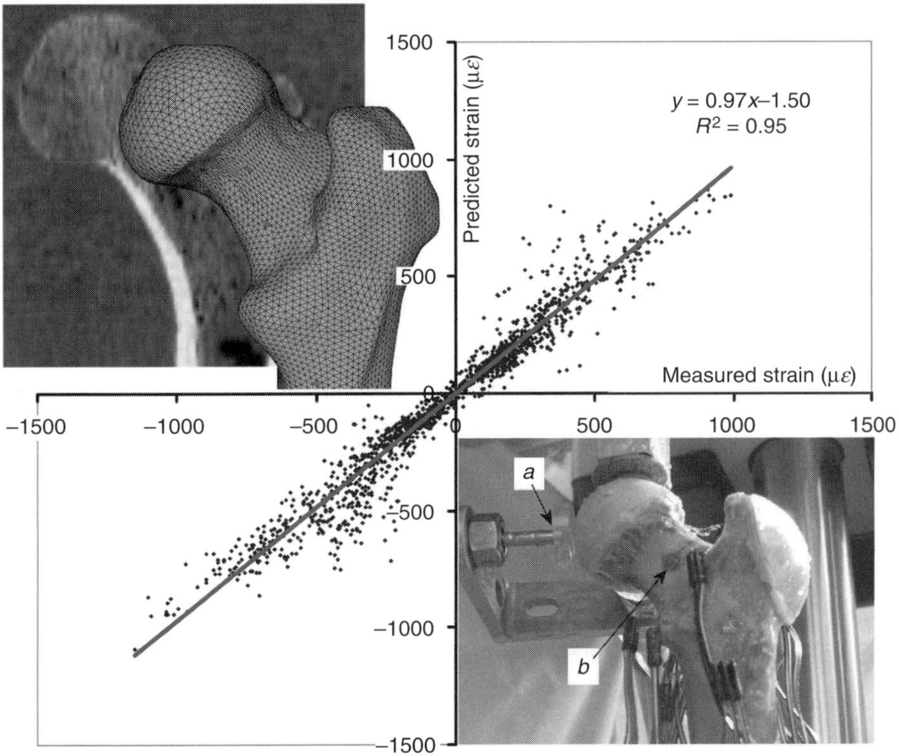

Figure 4.5 The finite-element model (top left) derived from a computed tomography dataset of one of the femurs used in the experimental validation study shows the position of the registered strain gages (b) applied on the specimen (bottom right). The LVDTs used to measure displacements in the proximal region are also visible (a). In the background, the regression plot between experimentally measured strains and model predictions.

observed during physiological activities. The CT data were processed according to the procedures described in the previous sections, and the resulting finite-element model was used to predict stresses, strains, and displacements at selected locations; by comparing these quantities with those measured experimentally on the same specimen, one could expect to validate the modeling procedure.

The careful reader has probably already spotted the trap hidden in this approach. From the CT data we derive two models: a structural model of the bone, and a heterogeneous constitutive equation for the bone tissue assumed as an elastic continuum. Unfortunately, both models require validation, and doing this jointly poses the risk that we overshoot one to compensate the deficiencies of the other. For example:

- I segmented the CT images with only moderate accuracy, and as a result some finite elements on the model surface fall partially outside the bone voxels in the image.
- I did not put the femur in water during the scan, so the voxels outside the bone have a very low density.

- I did not use special care in grading my mesh, and as a result the tetrahedral elements near the surface are fairly large.

As a result of these three methodological errors, the density assigned to the elements on the surface of my bone will be much lower than it should. Say that instead of 1.6 g cm^{-3}, we have 1.2 g cm^{-3} on average.

- I adopted the Carter and Hayes density–modulus constitutive law (Carter and Hayes, 1977), instead of the more accurate Morgan one (Morgan *et al.*, 2003). As a result, I ended up with a modulus of elasticity assigned to these superficial elements that is typically 4 GPa, instead of the more accurate 13–14 GPa that would be obtained with the Morgan equation and a lower partial volume effect. Notice that this error is limited to some elements on the surface of the femur, so overall this sequence of errors produce a reduction of the femur stiffness that is only, say, 30%.
- I adopted four-node tetrahedrons; these elements can only provide a linear interpolation of the local displacements. As a result my model result is stiffer than reality, say 30%.

If I use only strain and displacement measurements to validate my model I shall find a perfect agreement, even if the whole procedure is grossly wrong.

The solution is to use all available experimental data to uncouple these interdependencies. But first it is necessary to make a firm verification, and double check the model-to-experiment protocol, to be sure that predicted and measured quantities are compared at the same spatial location. For example, the error in the choice of finite-element type would probably be spotted with a simple mesh refinement convergence test, which would have showed that when using four-node tetrahedrons, the average element side must be really small, especially in the regions of high strain gradients.

Another possibility is to measure from the CT data the tissue density of the bone just underneath the strain gage, and use the modulus of elasticity derived from it, together with the plain strain hypothesis valid for any free surface, to calculate stresses from measured strains. Such "measured" stresses would then be compared with those predicted by the model. In the case given here, while strain would be predicted accurately, stresses would be markedly wrong, because in this case the error on the constitutive equation would not compensate that on the structural model.

Another way to validate such complex models is to use one controlled experiment to validate the model, and then use the model to predict other uncorrelated quantities that can also be measured experimentally. For example, in our case, we could use the model validated with the strain gage measurements to predict the load to fracture for that femur, and to compare it with the measured value. In our example, even if the predicted strains were very close to the measured ones, the model would make gross errors in predicting the location of the fracture initiation and of the load to failure.

5   Tissue level: mineralized extracellular matrix

A description of the various histological types of mineralized tissues that form bones, their biomechanics, and the approaches that are used to model such biomechanical behavior at the tissue scale.

5.1   Bone-tissue histology

5.1.1   Overview

In our journey from the organism to the molecules, we have now reached the tissue scale. Here the bone cannot be assumed as a continuum, as in Chapter 4, but we need to consider the *histology*, i.e. the microscopic anatomy of the tissue. Whereas gross anatomy describes organs as visible to the naked eye, so histological anatomy describes the structure and organization of the tissues that form such organs as observable with microscopes. Traditionally, histology involves: the fixation of a portion of tissue, so as to preserve it indefinitely; dehydration and infiltration with an embedding material, where dehydration is necessary because the water always present in the living tissues prevents the tissue from being infiltrated by the embedding material; sectioning, where the embedded tissue is sliced thinly, so that the slices become transparent to light; staining, which uses chemical reactions to color specific compounds and structures inside the tissue, and render them visible; and the microscope observation, which today can be done with a vast number of devices including scanning electron microscopy, laser confocal microscopy, etc.

Bone tissue is a composite material, made of a complex texture of *collagen* fibers that is mineralized by crystals of *hydroxylapatite*, also called hydroxyapatite, a naturally occurring mineral form of calcium apatite with the formula $Ca_{10}(PO_4)_6(OH)_2$ (Weiner and Traub, 1992).

A first histological discrimination involves noticing that specialized cells form an organic substance called *osteoid*, which later mineralizes into bone tissue. As bone is constantly destroyed and regenerated, in every portion of bone one can find tissue at different degrees of mineralization, including non-mineralized osteoid.

A second classification derives from the degree of organization of the collagen texture that forms the osteoid. In *woven* bone tissue, collagen is poorly organized, with short fibers oriented randomly; in *lamellar* bone the collagen texture is instead organized into repeating patterns. At the separation of two adjacent lamellae, there is usually a layer of small cavities, called *lacunae*, containing cells called *osteocytes*;

indeed, the term lamella (the diminutive of "lama," which means blade, which in turn derives from the Latin "lamina," which means "thin plate") refers to the layer of matrix between two rows of lacunae. Lamellae are usually a few micrometers thick.

A third classification is related to how the lamellae are organized to form the bone tissue (Rubin and Jasiuk, 2005). At the outer surface of bones, bone tissue is compact, with very little porosity; the lamellae are tightly packed to form shells of one or two millimeters. In regions where the thickness of compact bone becomes greater than a couple of millimeters, lamellae tend to fold around blood vessels, forming *osteons*, highly organized structures where the lamellae are organized with cylindrical symmetry one inside the other. In the internal regions of bones, where the porosity of bone becomes higher, lamellar bone forms a three-dimensional fabric of rods and plates, called *trabeculae*, to create a lightweight open-cell porous medium. Thus, lamellar bone can be further subdivided into *compact* and *trabecular* bone, and compact bone can also be *osteonal*. Of course, as usual in these cases there are a large number of synonyms: trabecular bone is also called cancellous or spongy bone, or even spongiosa. Osteonal bone is also called Haversian bone. Compact bone is sometimes called lamellar, but osteonal and trabecular are also lamellar.

5.1.2   Osteonal bone tissue

Osteonal compact bone can be found in most mammals, birds, reptiles, and amphibians. It is formed by many osteons organized in a parallel fashion interconnected by interstitial cement, and by interstitial lamellae, which are the remnants of osteons that were partially resorbed during the process of bone remodeling (Figure 5.1). Osteons are probably the strongest microstructural element of the human body, especially when subjected to compressive loads. The significant stiffness is due to the minimal degree of porosity, and also the peculiar orientation of the collagen fibers.

In textbooks, collagen orientation in osteons is described as follows (Figure 5.2) (Weiner *et al.*, 1997). Within each lamella forming an osteon, the collagen fibers are organized in a parallel fashion. However, in the most superficial lamella the parallel-running fibers are usually oblique with respect to the osteon's main axis (if we assume that the osteon is a cylinder, the main axis would be the cylinder axis). In the second lamella, wrapped inside the first one, the fibers are also oblique, but with an opposite orientation. This structure of concentric lamellae, with alternate orientation of the collagen fiber significantly increases the stiffness and the strength of the osteon to compressive loads.

As usual, the reality is a bit more complex than our textbooks describe it. One should never forget that anatomy (and histology for that matter) in its traditional sense is primarily intended to provide a descriptive, qualitative representation, not a quantitative one. If we pretend that these descriptions are detailed and quantitative, we face the risk of committing gross errors. As we proceed with a new, more physics-based approach to biomedical sciences, it is necessary to revise all those anatomical

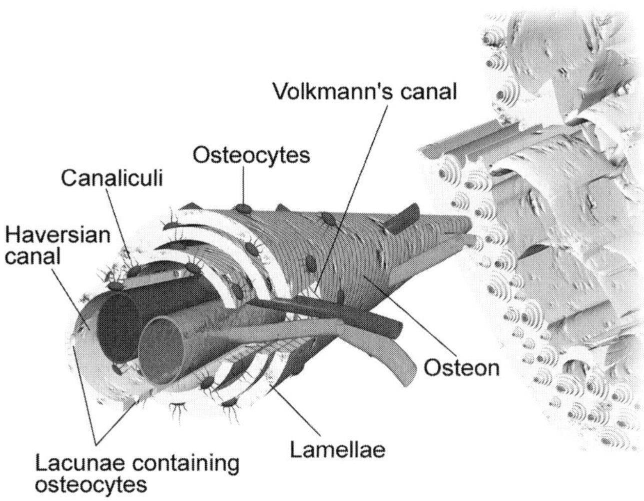

Figure 5.1 Pictorial representation of the osteonal compact bone tissue. The osteon has a central canal (Haversian canal) that usually hosts vasculature and innervations; from it depart thinner canals (Volkmann's canals) that connect to other Haversian canals; around it we notice the lamellae of mineralized collagen, nested radially to form a cylinder with the canal as the main axis. In between lamellae we observe the osteocytarian lacunae.

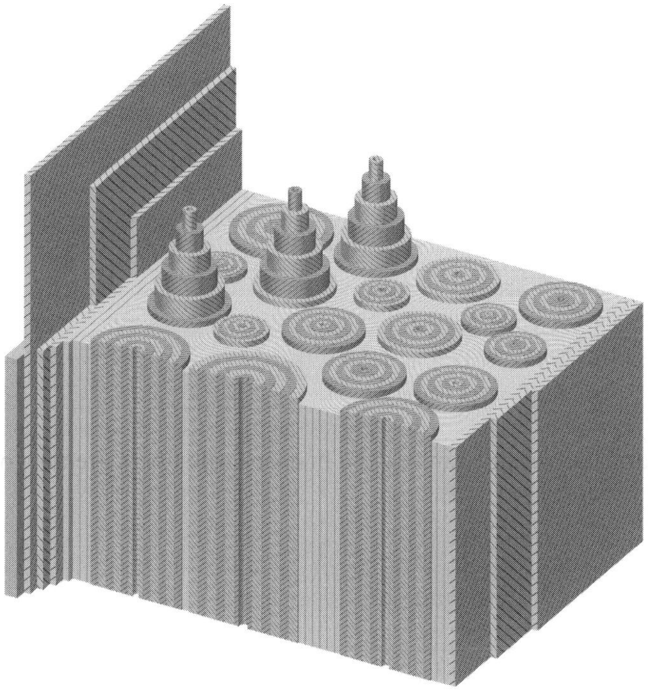

Figure 5.2 Spatial organization of collagen fibers in osteon lamellae, as described in textbooks.

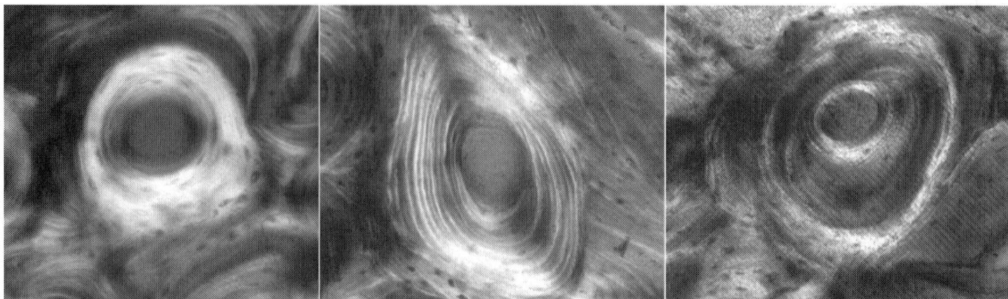

Figure 5.3   Circularly polarized luminescence (CPL) microscopy images. Left: bright osteons primarily contain collagen fibers oriented transversally to the osteon main axis. Middle: osteons with alternating dark and bright lamellae have successive lamellae with alternating orientations of the collagen fibers. Right: dark and hooped osteons contain lamellae with a predominantly longitudinal orientation of collagen fibers, except for most external and most internal lamellae, whose collagen is oriented transversally.

and histological observations so as to ensure that they provide the detailed and quantitative descriptions that activities like mathematical modeling require.

Collagen orientation in osteon is not an exception. Recent studies on long bones have highlighted how in reality collagen can be oriented in very many ways in osteons (Figure 5.3) (Skedros *et al.*, 2009), and this orientation, while random in nature, shows a correlation with the type of bone and the location in the bone length (Beraudi *et al.*, 2010). This observation might suggest a third mechanism of skeletal adaptation to loading: not only do bones adapt their shape and degree of mineralization and porosity to the loading environment, but they might also exhibit different patterns of collagen orientation in osteons in the various regions of a single bone, most likely depending on the type of stress history (tensile, compressive, shear) acting in that region.

Another element of great interest in osteonal histology is the fluid circulation. In most osteons the central cavity, called the *Haversian canal*, is occupied by blood vessels and nerves. *Volkmann's canals*, which connect Haversian canals to each other and to the periosteum, provide another channel for fluid. A third level of porosity is provided by the *canaliculi*. As mentioned before, there is usually a layer of small cavities, called *lacunae*, at the separation of two adjacent lamellae containing *osteocytes* (Ardizzoni, 2001). These lacunae are interconnected by the canaliculi, small channels that perforate the bone lamina and provide a space where the osteocytes can extrude their long processes. Haversian canals, Volkmann's canals, and canaliculi are all interconnected with each other and the periosteum (Cooper *et al.*, 2003), ensuring that all regions of bone tissue, even the most compact ones, are permeated with fluids that transport oxygen and nutrients, and remove metabolites. All the important bone biology processes (apposition, resorption, fracture healing) require that large amounts of cell precursors are deployed where necessary within the bone tissue, and are kept vital while they differentiate and fulfil their interaction with the extracellular matrix. This implies that bone tissue, even osteonal tissue, must have appropriate macro- and micro-circulation. Owing to the tortuosity of this system of large and small channels, the permeability of cortical bone is quite low, and it seems unlikely that micro-circulation can be ensured only by the cardiac output. However, when bones are loaded, the fluid trapped into these channels is squeezed and accelerated; the loading is cyclic, this provides an additional pumping mechanism that ensures that all regions of the skeleton are properly perfused.

Osteons and osteonal remodeling are strongly affected by various factors, and the resulting structure somehow contains the history of the organism. Thus, osteons can provide hints on the sex, health history, lifestyle, or dietary regime of individuals. Osteon size and arrangement varies according to taxon (group of organisms, generic name for the elements of a taxonomy), so that genus (lower level taxon) and sometime species (lowest level taxon) can be differentiated from even a few bone fragments, which is sometime necessary in forensics and archeology.

While most osteons run according to preferential orientation (i.e. parallel to the bone long axis in long bones) some osteons "drift" in other directions, or change direction along their length (Robling and Stout, 1999). These are called drifting osteons, and their existence is not yet fully explained.

5.1.3 Lamellar compact bone

A layer of lamellar compact bone wraps all cancellous bone regions with thicknesses of 0.5–3 mm. Although to my knowledge this has never been systematically investigated, I suspect that there is a characteristic thickness above which lamellar bone start to curl into osteons; this could also explain why in very small mammals, like the mouse, osteons are rare.

5.1.4 Cement

As mentioned already, bone tissue also includes an amorphous material called *cement*, composed mostly of glycoproteins, mucopolysaccharides, lipids, carbonate, and citrate. The other constituents include sodium, magnesium, and fluoride. Cement shows viscous and plastic behavior, and forms the interstitial cement line that marks a previous remodeling cycle, as well as the ground substance surrounding osteons.

5.1.5 Cancellous bone

Cancellous bone has a very complex morphology that varies considerably depending on anatomical site and age, and also in relation to pathologies, such as osteoporosis where bone metabolism is altered. Cancellous bone is less stiff than compact bone and somewhat weaker, although the stiffness and strength of cancellous bone can vary considerably depending on its porosity (usually called the bone volume fraction, i.e. the fraction of the tissue volume that is occupied by mineralized matrix), its mineral content, its spatial organization, and probably also on the organization of the collagen that forms the tissue.

Cancellous bone has a higher surface area than compact bone, is highly vascular, and its porosities are frequently filled with red bone marrow, a tissue with a high concentration of undifferentiated cells, in particular hematopoietic stem cells capable of differentiating into all blood cell types. Thus, it is not surprising to find that cancellous bone is more prone to adaptation and remodeling, and tends to resorb more quickly than compact bone in osteoporotic patients. Even in normal physiological conditions, cancellous bone is riddled with remodeling sites, where old bone lamellae are resorbed and new bone is apposed in the form of osteoid, which later mineralizes.

In general, it is believed that at the lamella level compact and cancellous bone are identical both in term of collagen organization and of mineralization. However, the presence of so many remodeling sites makes it quite possible to find trabeculae with a region that is only partially mineralized. If we measure the degree of mineralization of a portion of cancellous bone, we notice a distribution of values; it has been written that this distribution has a characteristic shape under physiological conditions, and when it changes this is a sign that the remodeling activity is somehow altered (Roschger *et al.*, 1998; Fratzl-Zelman *et al.*, 2009).

5.2   Bone-tissue biomechanics

In Chapter 4, I showed that we may assume whole bones as continua, obtaining fairly accurate predictions of stresses and strains. When we move down to the tissue scale, however, this assumption cannot be accepted anymore, as the size of the porosities is comparable to the characteristic dimension of the structure we are investigating. Now we assume that the mineralized material that composes the tissue is a continuum, but that the tissue itself is modeled as a structure.

Another peculiar aspect of the tissue scale is that it is the scale where the interaction between the biomechanical function and the biological form becomes most evident. Today, this is the preferential scale at which mechanobiological processes are modeled and described.

In this section, I shall describe the biomechanics of bone tissue, reporting the primary experimental results collected so far. In the following sections, I shall report on how the tissue can be modeled to predict its elastic behaviors and its failure, and in relation to the modeling of mechanobiological processes.

5.2.1   Factors affecting the mechanical properties of bone tissue

The mechanical behavior of bone tissue has attracted the interest of many researchers in the past, but the one who probably established the largest body of empirical knowledge on this argument is Gaynor Evans. In his fundamental books, Evans reported the results of dozens of experiments where bone was investigated as a material (Evans, 1957; 1973a); see also (Evans, 1973b).

Firstly, Evans faces the issue of whether testing cadaveric bone would yield results representative of the properties that the same bone tissue would have in vivo. By systematically revising the relevant literature available at that time Evans showed how many of the alleged differences were indeed due to the effects of drying and preservation on the mechanical properties of bone tissue, and that when this effect was removed, no significant differences were observed between measuring the mechanical properties of bone tissue in vivo or ex vivo.

With respect to the water content, Evans showed, conclusively, how significantly the dehydration affected the mechanical properties of bone tissues; it is now recognized that any biomechanical test on bone tissue must be designed to retain the tissue moisture at all times.

With respect to preservation, Evans was clear in advising against the use of alcohol fixation, and in favor of the use of deep freezing (-20 °C). The results on the effect of embalming were less clear. A recent study (Ohman *et al.*, 2008) has clarified that there is another parameter to be considered, and this is the duration of the embalming bath. Bone tissues preserved in an aqueous formalin solution of 4% for four weeks or less show no significant differences in their mechanical properties when compared to fresh tissues. However, if the tissues are kept for eight weeks in the embalming fluid, significant differences appear. It should be noted that today regulations impose extreme caution when potentially infected tissues are subjected to conditions that could potentially project them at high speed into the environment, as is always the case when biomechanical tests are involved. While deep freezing is not considered a safeguard against biological hazards, formalin is known to kill virtually any virus, bacterium, etc. In this sense, handling tissue specimens fixed in formalin should be safer (on the other hand, formalin is itself a hazardous substance, the handling of which must made with great care).

Another determinant when measuring the biomechanical properties of bone tissue is the tissue temperature. Ideally, all tests should be at 37 °C, but in a number of experimental set-ups this is difficult to realize. The elastic properties of bone tissue tend to change linearly with temperature: the module of elasticity if measured at 23 °C would differ by 2–4% from that measured at 37 °C (Cowin, 1988). Thus it is reasonable to perform these tests at room temperature, so long as this is monitored and eventual fluctuations are accounted for. Other properties seem to be more sensitive (fatigue strength, toughness, etc.) and the specific methodological literature should be investigated before such tests are designed.

Bone mechanical behavior is well described by viscoelasticity; thus, it is no surprise that bone-tissue properties are in general sensitive to the duration, frequency, and rate of loading. Without attempting an exhaustive summary of what is a very extensive body of literature, we can say that when bone is subjected to non-strenuous physiological conditions, such as those produced by slow walking, stair climbing, etc., the error we commit in modeling bone tissue as a purely elastic material is very small; however, when the bone-tissue properties need to be investigated in conditions typical of strenuous physiological (running, jumping) or para-physiological (falling, projectile impacts) conditions, such simplification is no longer acceptable.

Also, particular care should be given when designing experimental measurements of bone-tissue properties to creep or relaxation phenomena. Bone does creep at room temperature, and if this aspect is not taken into account, it can become a source of considerable experimental error.

There are dramatic differences in the biomechanical properties of bone tissues of different species, even if we limit ourselves to mammals. The tensile module of elasticity of the femoral cortical bone is 18 GPa in humans, 25 GPa in horses and cows, and only 15 GPa in pigs (Fung, 1993). Thus, caution should be taken in using properties derived from animal tissues in the modeling of human bones.

Bone-tissue properties change significantly with age. Children have softer and tougher bones, whereas the elderly tend to have more fragile bones. Also, there is progressive loss of bone mineral density with age. However, the biomechanical properties of bone tissue are quite similar in individuals aged 20 to 50 years old.

Sex in itself has no major effect on the biomechanical properties of bone tissue. Men tend to be bigger and heavier than women, and thus their bone mass tends to be bigger, but the tissue forming their skeletons is not different in quality. Significant differences begin to appear after menopause, but this is because in women the progressive loss of mineral density is faster than in men, probably in relation to the hormonal changes induced by the menopause.

With respect to phenotype,[1] all observations in this sense are necessarily inconclusive. Human phenotypes tend to cluster geographically, and these groups share similar lifestyles, diets, climate, etc. Thus, it is practically impossible to know if any of the differences we might observe are truly due to genetic traits or to these complex environmental and cultural determinants.

The discussion on what Evans call "topographic differences" is also complex. If we look at whole bones, we shall see significant differences in their biomechanical properties, agreeing with the generally accepted theory that the skeleton tends to adapt globally and locally to the specific functional conditions it has to face. So the radius, which is subject to the most variable loads, is the usually the strongest in bending, whereas the femur, which is almost always loaded as a column, is the strongest in compression. But even assuming there is a causal relation between form (intended here as properties) and function, at which dimensional scale does the bone adapt to the function? Certainly, different bones have different shapes, and different bones have very different distributions of the various histological types (some bones even lack one histological type). Various authors have found that the same histological type of tissue

---

[1] I do not recognize any scientific foundation in the concept of race, and refuse to use this term in this book. The human phenotype varies continuously, although we can cluster groups of people by using similarity concepts on certain traits. So if the group "black" is based on the color of the skin, many people from central Africa would be included into it, but depending on the tone we use to cluster, also many from north Africa, southern Italy, many people from Turkey, etc. The need to cluster people into groups is a cultural artifact, usually motivated by cultural differences rather than by physical differences. However, the need to call these differences "racial" and the need to show that such "racial" differences have a genetic basis has always been motivated in human history by scopes and purposes with which I disagree. Since the Human Genome Project has finally proved conclusively that there is no genetic basis to the concept of race, I refuse to use this term in my research activity and invite my colleagues to do the same.

shows different morphological and biomechanical properties in different regions of the same bone. As mentioned, it has been shown how even the orientation of the collagen fibers in osteons varies along the length of the diaphysis of long bones (Beraudi *et al.*, 2010). These observations would suggest that bone tissue adapts to functional conditions by changing its chemical composition, the organization of the molecules, the organization of the texture, the porosity of the tissue, then the thickness and the ratio of histological types of the tissue, and the shape of the whole bone. As discussed in another section, I reject the idea that the skeleton is an optimized structure (optimal for what?) but I recognize that the skeleton is an *adaptive* structure; what I suspect is that this adaptation happens at several dimensional scales simultaneously from the molecules to the whole structure, in a way that is not only difficult to model but even to conceive! But apart from this aspect, we can definitely conclude that the topographic effects on the biomechanical properties of bone tissue are important.

5.2.2 Constraining, loading, and tissue orientation

From the previous section it should become quite clear that the biomechanical properties of bone tissue are affected by a very large number of determinants, some of which can be controlled by the experimenter with some effort, others which are very difficult to control. But as we shall see in this section even the concept of measuring bone-tissue properties is a delicate one.

A first general problem is that the morphological complexity and the porosity of bone tissue always create problems when a portion of tissue is isolated and tested. Consider a portion of cancellous bone. To obtain this portion, some trabeculae had to be resected. While that portion of the tissue was *in situ*, those trabeculae were connected to the surrounding tissue, providing a general stiffening of the structure. When we test the specimen instead, these superficial trabeculae are unconnected, and thus their contribution to the structural stiffness is much lower. The same reasoning can be applied to osteonal bone, when by taking a tissue sample we cut some osteons on the surface; it is clear that the stiffness of half an osteon is not even comparable to that of a whole osteon. This makes any superficial measurement questionable. For example, the data on elongation–shortening, usually obtained from surface-attached extensometers are traditionally unreliable, and it is recommended to mount four of them, one on each side of the specimen, so as to average out random errors (Keaveny *et al.*, 1997).

A second problem is related to the alignment of tissue specimen with respect to the predominant orientation of the tissue, and then the orientation of the specimen with respect to the loading direction. If the tissue presents a predominant direction (e.g., the cancellous bone tissue in the femoral neck) special care has to be taken over how the specimen is cut with respect to this predominant tissue direction and, even more importantly, how such alignment is measured on each specimen (Ohman *et al.*, 2007). In a recent paper, my colleagues and I were able to explain 95% of the variability of the stress to failure over a large pool of osteoarthritis bone tissue samples with an inductive model that considered the mineralized volume fraction, the orientation of the specimen with respect to the predominant tissue orientation, and the principal eigenvalues of the fabric tensor, which accounts for tissue anisotropic orientation (Tassani *et al.*, 2010). This suggests that both the directions (that of the tissue and that of the loading) should be considered and accounted for.

5.2.3   Specimen size and shape: the continuum assumption

The collagen fibers in the osteons are strongly oriented in space; thus it should not be surprising to discover that if we test osteonal compact bone, or even single isolated osteons, we shall find significant difference between the biomechanical properties measured along the axis parallel to the predominant orientation of the osteons, or measured in any direction orthogonal to this one. When tested along the osteons' alignment, osteonal bone is much stiffer and stronger; the module of elasticity is almost double in the axial direction, with respect to any off-axis direction. However, if we create an accurate finite-element model of the whole bone, and we compare the predictions obtained assuming cortical bone to be isotropic or anisotropic, very small differences appear between the strain fields (Baca *et al.*, 2008; Peng *et al.*, 2006; Yosibash *et al.*, 2010). Why? This seems to contradict common sense, and cast doubt on the accuracy of numerical models. However, if we use the same model to predict the strains we measure experimentally, we shall find excellent agreement (Schileo *et al.*, 2007). The explanation of this apparent contradiction is that bones are structures; while in isolation a single osteon, or a small portion of cortical bone, is found to be highly anisotropic, the shape and composition of whole bones tend to obscure these differences. So long bones, made of slender tubes of dense cortical bones, tend to deform axially, while circumferential and radial deformations are much smaller. Thus, even if I reduce by 50% the module of elasticity in the circumferential and radial directions, the deformation of the structure will be dominated by the axial stiffness.

Steel has a crystalline structure; the crystallites that form the steel are called grains. In steel, a grain size is typically 10–20 micrometers and a specimen to measure the mechanical properties of steel is typically 20 mm wide (10^3 times the grain size). Thus when we make a specimen of steel some 10 mm wide, we are quite sure that what we shall measure is the average properties of very many grains, and such properties will be quite insensitive to the local features of single grains. By comparison, an osteon is 200 micrometers thick and 1000 micrometers long. In human beings the thickest cortical wall can be 10 mm, but owing to the curvature flat specimens are rarely thicker than 2–5 mm (10 times the osteon size), and sometimes much less; the problem is also the same for cancellous bone.

The first to raise this problem, with cancellous bone, was Tim Harrigan in his 1988 paper (Harrigan *et al.*, 1988), where he concluded, "Within three to five trabeculae from an interface a continuum model is suspect." Tony Keaveny, in his seminal 1997 paper (Keaveny *et al.*, 1997), argued quite convincingly that all previous studies where cancellous bone was tested using platen compression tests were affected by systematic errors of 20–40%, and random errors of up to 12%. More recently, our group showed that by using cube-shaped specimens instead of cylindrical specimens in cortical bone compressions tests, systematic errors of up to 60% were obtained (Öhman *et al.*, 2009).

The general evidence that emerges from all these studies is that testing bone tissue as a material, i.e. under the continuum assumption, is a challenging operation, which might or might not yield acceptable levels of accuracy, depending on a large number of factors.

This is why, in parallel with a progressive refinement of the experimental methods, researchers have developed other approaches, where the tissue specimen is recognized as a structure, and not as a continuum, and the constituent material properties are derived by solving an inverse problem (usually by means of a numerical model coupled to the experimental set-up) that predicts these properties from the measured quantities. This approach will be discussed in full in the following section.

5.2.4 Tissue properties through the inverse problem

Before we begin the argument, it is fair to mention that in principle it is nowadays possible to use micromechanical tests to measure directly the properties of the constituent biological materials that form the bone tissue. Atomic force microscopes, nano-indentation, etc., make it possible to measure the module of elasticity of a single trabecula, or a single osteon. However, while the complexity of these experiments is significantly higher, moving down one scale in space does not solve the problem. The average thickness of a bone trabecula is 100–200 micrometers, and the osteocyte lacunae that permeate it are typically 10–20 micrometers in diameter; thus we are still not that far from the ten-to-one ratio between specimen size and the size of the structural element of the material, which was considered a problem for cortical bone specimens in the previous section. Thus, while these nano-mechanical methods are of great value in disclosing information at a much smaller scale, they do not in themselves solve the problem here under discussion.

So if we cannot assume that the specimen under testing is a continuum, and derive "material" properties almost directly from measurements based on this assumption, then we need to accept that the specimen is a structure and that the relationship between what we can measure and the tissue properties is complex. However, the relationship can be determined using inverse methods, where a detailed numerical model of the specimen is used to establish the relationship between the quantities we can measure and the true tissue properties. Talking about tissue, there are two possible inverse problems. We can model whole-bone experiments, and use them to validate general tissue empirical equations obtained experimentally. Or we can model the tissue specimen and used it to invert the measurements on the specimen into tissue properties. The first example has been discussed in detail at the end of Chapter 4. Here, I discuss the second case, relevant at the tissue level.

The tissue specimen is scanned using micro computed tomography (microCT). These laboratory systems can typically accommodate specimens of 20 mm or smaller, but can produce images with a spatial resolution of 10 micrometers or better. Using the methods described in the following section, it is possible to convert these images into a finite-element model of the mineralized portion of the specimen, the only portion that exhibits appreciable mechanical properties. The experiment (usually a compression test) is simulated, and the tissue-level module of elasticity is varied in the numerical model until optimal agreement with the measured force–displacement curve is reached.

It is evident that this inverse solution is obtained under the assumption that the mineralized extracellular matrix that forms bone has constant mechanical properties. This assumption that bone matrix is a homogeneous continuum is much debated (Tassani *et al.*, 2011), but if we use the method proposed, and thus model only the portion of bone matrix that is fully mineralized, such an assumption appears to be acceptable in most cases.

It should be noted that when a tissue-level model is identified in the matrix properties via an inverse modeling process like the one described here, the same set of experimental measurements cannot be used to validate the model. We need to produce a whole new set of *independent* measurements, which is not trivial. One example could be to identify the constitutive equation of the bone matrix via inverse modeling, and then use the identified model of the tissue specimen to predict its load to failure.

5.2.5   Direct measurement of tissue strain

So far, we have not considered the simplest option: if we can measure the tissue strain induced at various locations of the specimen by a known loading condition we can much better understand the tissue biomechanics, and also produce data that are essential in the validation of numerical models predicting the bone-tissue biomechanics. The problem of measuring the strain of a structure as complicated as cancellous bone at a scale of 10–100 micrometers is complex, and beyond the scope of this book. Over the years, researchers have explored the use of X-ray diffraction, ultrasounds, optical sensors, etc. Recently digital image correlation methods, based on both optical and X-ray imaging, have shown encouraging results. However, to date the problem should be considered still open, and thus a target of further methodological research.

5.3   Modeling bone-tissue biomechanics

If we represent bone tissue as a porous structure formed by the mineralized matrix, and the rest of the volume occupied by other histological types (osteoid, marrow, etc.), predicting the deformation of such tissue under the action of external forces remains a continuum solid mechanical problem, which is best solved by numerical methods, such as the finite-element method. In this respect, modeling bone at the tissue level could be seen to be the same problem as modeling bone at the organ level. But there are a few fundamental differences.

5.3.1 Imaging sources

While modeling bones at organ level is possible starting from clinical CT data of the bone of interest, for tissue we need to use an imaging technique with a much higher resolution. Clinical CT scanners have a spatial resolution that in most ideal conditions is 300 micrometers, clearly insufficient to see bone trabeculae, the average thickness of which is 100 micrometers. High-resolution peripheral quantitative CT systems do better, with a typical resolution of around 80 micrometers, but the radiation dose that would be delivered makes them applicable only to image distal limbs (Boutroy *et al.*, 2005). Recently high-resolution imaging methods have been proposed that could in principle be used to image tissue in the axial skeleton, with a resolution of approximately 150–200 micrometers (Phan *et al.*, 2010). Currently, the only option to achieve the resolution of 20–40 micrometers necessary to have an accurate morphology of cancellous bone is using microCT machines for ex vivo imaging, which produces detailed images of bone biopsies. A recent study showed that the accuracy of non-linear microscopic finite-element models in predicting the yield load of cancellous bone specimens in compression is sensitive to the resolution of the microCT images used to generate the model (Bevill *et al.*, 2009). The authors found that these errors are acceptable up to 80 micrometers, but with image resolutions of 120 micrometers the errors can be as big as 80% in highly osteoporotic tissues. The problem with this kind of study is that the results they obtain are of course valid only for the specific type of modeling procedure used. As we shall see, there are modeling methods that are less sensitive to the topological accuracy than others, so we cannot exclude that in the near future a combination of advanced imaging and robust modeling methods might make in-vivo tissue image-based modeling possible everywhere: but today the reality is that we can do this only at the wrist and the ankle, whereas the rest of the skeleton can be analyzed only through biopsies.

5.3.2 Modeling methods

As mentioned, the biggest issue with biomechanical modeling of bone tissue is the exceedingly complex topology of the domain of integration. This makes the segmentation of the 3D image with topologically correct polygonal surfaces difficult, but the automatic mesh generation of the resulting solid is even more challenging, owing to the presence of thin rods and plates.

For a critical review of all available methods for automatic mesh generation in the specific application domain of image-based bone modeling, see (Viceconti and Taddei, 2003). Here, I shall only briefly touch the principal points.

All classic automatic mesh generation algorithms (Delaunay, wavefront, etc.) work only if the average element size is set to values so small that the resulting mesh would have billions of elements. There are two possible solutions, both drawing on the fact that the initial information (the microCT volume) is a regular 3D lattice: Cartesian meshes and meshless methods.

Cartesian meshes are usually referred to in the biomechanics literature as *voxel meshes*, because they are generated by directly converting each voxel in a 3D image into a cubic finite element. The mesh generation becomes trivial as well as the image density mapping (each voxel element has a single density value). The boundary of the domain of integration is represented by a jagged surface, which tends to the true boundary as the voxel size tends to zero. The only way to achieve a reasonable accuracy with voxel meshes is to have finite elements as small as possible. In tissue-level imaging, where microCT machines are used, we can have a billion voxels. If we keep this as the element size, the jagging becomes relative moderate. Of course, this implies solving a finite-element model with some hundred million degrees of freedom. In Chapter 2 I explained that a typical sparse solver requires $O(NB^2)$ operations. Assuming 90% sparseness, a problem with 300 million degrees of freedom would have a bandwidth around 25 000; it would require a 1 Gflops processor almost six years to solve this problem, and more memory than is currently available on any computer. While computing speed can be improved significantly, the memory occupation remains a serious bottleneck, and thus for very large problems it is usually preferable to use iterative solvers, where clever pre-conditioning methods can drastically reduce both execution time and memory footprint.

Such pre-conditioners can become particularly efficient if, as in our case, all finite elements in the model are identical in shape and size. Two methods were originally proposed in this context: row-by-row (RBR) and element-by-element (EBE). The first is much more efficient, but requires that all elements are not only identical in shape, but also in their material properties (Van Rietbergen *et al.*, 1996); on the contrary, the EBE approach allows the use of heterogeneous material mappings, which is known to improve the model accuracy (Homminga *et al.*, 2001). When combined with state-of-the art pre-conditioned conjugate gradient (PCG) iterative solvers, these pre-conditioners are very memory efficient (the stiffness matrix is not explicitly constructed), and the work per iteration, as well as per degree of freedom, is constant. However, the number of iterations required to reduce the residual by a constant fraction rises dramatically as the problem size increases. To address this problem, and to take advantage of modern massively parallel super-computer architectures, algebraic multigrid method (AMG) pre-conditioners were introduced (Adams *et al.*, 2004). Using the ASCI White IBM SP Power3 super-computer Adams and colleagues were capable of solving a linear model of 500 million degrees of freedom in approximately one hour. Using an IBM *Blue Gene* BG/L Supercomputer and the ParFE solver, a similar linear problem with 500 million degrees of freedom was solved in 6.2 minutes, and a linear model with 1.5 billion degrees of freedom was run in half an hour (Bekas *et al.*, 2008).

Meshless methods are more recent, and their scalability to very large problems remains to be proved, although in principle many of the speed-up strategies mentioned for the Cartesian meshes could also be applied to meshless methods. The advantage of meshless methods is that the model can be initialized with a cloud of points, and progressive refinement schemes, such as the octree approach, can be adopted in a straightforward manner, since with meshless methods it is not

Figure 5.4 Output of the MCM model of a small cube of cancellous bone tissue imaged with a microCT system at a resolution of 40 micrometers. Shading represents the intensity of the displacement vector.

necessary to deal with the problem of dangling nodes that affects octree refinement schemes of finite-element meshes. Among meshless methods, the *meshless-cell method* (MCM) described in Chapter 2 appears to be particularly interesting. For high-resolution models, MCM produces results almost identical to those of finite-element models (Figure 5.4), but for lower resolutions, the accuracy degrades much more slowly. This means that it is possible to achieve the same level of accuracy provided by voxel meshes with a much coarser model, and thus much smaller computational problem. While this does not seem to be a problem for single-run models, such lighter computational weight could be a real bonus for cases when the model has to run many times, as part of Monte-Carlo probabilistic schemes, for non-linear models, or when the model is linked to a bone remodeling algorithm.

5.3.3 Identification of the constitutive equations

The aforementioned difficulties in measuring strain at the tissue scale can make quite captious the discussion on the most adequate constitutive equation to be used to predict stresses and strains inside bone tissue.

The simplest approach used in the literature for cancellous bone tissue involves the inverse determination of an average modulus of elasticity for the mineralized extracellular matrix from the load displacement curve obtained from a compression test on the tissue specimen. In this case, it is assumed that the bone matrix is an isotropic and homogeneous material.

A second method uses the very same approach adopted in the conversion of CT data into finite-element models at the organ level: the microCT data are calibrated against mineral density using a calibration phantom, and the matrix properties are described as heterogeneous and locally isotropic (heterogeneity produces large-range anisotropy), where the modulus of elasticity depends on the local mineral density of the tissue. This approach makes the model less sensitive to the level of thresholding used to segment the bone tissue; one can use relatively low thresholds, being sure that low-mineral regions will be properly accounted for in the model. The problem is the density–modulus empirical relationship, which is not necessarily the same we use at the organ level. A third approach involves the use of nano-indentation to estimate the elastic modulus of the bulk tissue (Lewis and Nyman, 2008).

The complex geometry of cancellous bone tissue might create stress increases high enough to bring small volumes of tissue beyond the elastic limit while the rest of the tissue behavior is still largely linear and proportional. If we neglect this phenomenon and pretend that the tissue is indefinitely elastic, significant errors can arise in the prediction. Since here the interest is not in modeling the behavior of the tissue under large plastic strains, but only in preventing microscopic islands of plasticity from deteriorating the accuracy of the model, a fairly simple elastic and perfectly plastic constitutive equation can be adopted, where the bone is assumed to behave perfectly elastically up to a given stress or strain value, and then to deform with perfect plasticity, i.e. deforming plastically without any need to increase the stress. Such a very simple post-elastic constitutive equation is fully identified when in addition to the elastic contacts we define the elastic limit, which is usually defined over the strain.

A second type of non-linearity appears if we want to take into account that, owing to their slender shape, some trabeculae, especially in low-volume-fraction tissue, might show significantly large displacements, large enough to induce significant errors if the assumption of small displacements is made. This, associated with large displacements, is the simplest structural non-linearity to treat; usually every code that allows the definition of a post-elastic constitutive equation also allows large displacements to be accounted for.

Some authors are working on simulations with much more complex phenomena, such as fracture or self-contact (a trabecula bends until it touches another trabecula); needless to say, trying to account for these phenomena dramatically complicates the model. However, to my knowledge no specific clinical question so far seems to require that we model the behavior of the bone tissue so far beyond the elastic limits.

5.3.4   Boundary conditions

So far, the problem of boundary conditions has been investigated in very limited ways. The primary concern has been to replicate as accurately as possible the boundary conditions applied to the tissue specimens during micromechanical testing. To my knowledge, no one has yet tried to account for the connectivity between the portion of tissue being modeled and the rest of the bone.

5.3.5   Verification, sensitivity, and validation

So far, most modeling work at tissue level has been done with modeling techniques that involve an extremely high computational cost. This has somehow limited the extent of methodological assessment.

To my knowledge, no extensive sensitivity analysis has been conducted on tissue-level models. Such a study would be particularly interesting with respect to certain modeling details, whose sensitivity of the final predictions remains unclear.

Also validation at tissue level is much more limited than at the organ level. In most cases, the validation is limited to the solution of the inverse problem; the prediction of the modulus of elasticity of the bulk material (the extracellular mineralized matrix) from the specimen force–displacement curve, as compared to that measured with nano-indentation. Of course, this information tells us only that the integral behavior of the model is consistent with what is measured, but does not ensure that the local predictions, such as displacement, stress, and strain, are in any way valid.

6   Cellular level: cell–matrix interaction

A description of the bone extracellular matrix from a biochemical point of view and the specialized cellular types that permeate it, as well as their interaction as part of the mechanobiological interaction called bone remodeling.

6.1   Bone extracellular matrix

In connective tissues, the extracellular matrix is the extracellular part of tissue that provides structural support to the cells. The major part of bone mass is made of the *bone extracellular matrix*. The bone matrix is made of an inorganic part and an organic part: it is composed of a protein matrix, which contains embedded crystals of hydroxyapatite, a form of calcium phosphate. Bone-forming cells secrete the organic part, called *osteoid*, formed by collagen and other noncollagenous proteins. The osteoid can be laid down quite rapidly without any particular spatial organization, which produces woven bone, or more slowly in a more organized fashion, which produces lamellar bone. Once the osteoid is apposed, it starts to mineralize; inorganic materials precipitate on it, stiffening the tissue considerably.

6.1.1   Collagen

Approximately 90% of the weight of the bone organic matrix is collagen. This protein is very common in the human body (nearly a third of the whole protein mass). There are many different chemical types of collagen in the human body, but the most common in bones is type I; to which I shall refer in the following.

Collagen is secreted by cells in the form of tropocollagen, the base unit that forms larger collagen aggregates, such as fibrils; somewhat improperly, tropocollagen is called the "monomer of collagen." The tropocollagen molecule, around 1–2 nm in diameter, and typically 300 nm long, is formed by three polypeptide chains, each spatially organized as a left-handed helix; these three chains are twisted together to form a right-handed coil stabilized by various hydrogen bonds. A polypeptide is a natural polymer that links multiple α-amino acids; in collagen, the most common are glycine (G), proline (P), and hydroxyproline (Hp). These are most commonly organized in repetitive patterns, either G-P-X or G-Y-Hp, where X and Y are amino acid residues other than G, P, or Hp (Behari, 2009).

These regular patterns are responsible for the helical structure of the molecule, and its tendency to link with other molecules of the same kind to form larger conglomerates. As it is secreted into the extracellular space, tropocollagen starts to self-assemble into large macromolecules using a staggered pattern that leaves gaps in the structure of the order of 40 nm. Also, the macromolecules have an elongated aspect ratio – the word *fibrils* – which organize into bundles to form collagen *fibers* (Silver, 2009).

The determination of the mechanical properties of fundamental constituents of living tissue, such as collagen, is very difficult (Strasser *et al.*, 2007). Collagen fibers are very small and their testing requires very complex nano-mechanics methods, the accuracy of which is not always fully documented. Being so close to the molecular scale makes the continuum assumptions used in classical mechanics of materials quite delicate, and frequently we observe very large variations from one study to another, most likely because some molecular conditions, such as the level of hydration, are different. This is currently (2009–2011) an area of intense research and it is possible that further development may radically change what is reported here. Having said this, we can say that type-I collagen fibers are capable of sustaining only tensile loads, but under such loads they express a considerable stiffness: the module of elasticity is 0.8–8.0 GPa, depending on the level of hydration, the size of fibers under testing, etc. (Eppell *et al.*, 2006; Shen *et al.*, 2008; Yang *et al.*, 2008; Yoon and Cowin, 2008). Among constituents of connective extracellular matrices, only hydroxyapatite is stiffer. Under tensile loading, collagen type-I fibers elongate quite linearly with load, except for an initial non-linear toe region owing to the uncoiling of the molecules, which however, is much less marked than in other fibrous proteins, such as *elastin*. The post-elastic behavior of collagen is much less clear: in some cases, a fragile fracture is observed around a strain of 30%, whereas in many others the fibers start to deviate from linearity around 10–20% of strain, showing a sort of yield behavior with transitions from regions of strain hardening to regions of strain softening (Roeder *et al.*, 2002; Shen *et al.*, 2008). Some molecular dynamics studies suggest that this might be because of the aspect ratio between the fiber diameter and the size of the defect, which might or might not become critical. But this discussion is somehow academic in the context of this book: the elastic limit of bone tissue is around 1%, so the mineralized phase will start to fail macroscopically well before the deformation is large enough for the collagen behavior to deviate from linearity (Tang *et al.*, 2010).

6.1.2　Non-collagenous proteins

In addition to collagen, there is long list of non-collagenous proteins (NCPs) that can be found in bone tissue, some specific to this tissue, others also present in other connective tissues. Although in mass they contribute only a few percent, osteoblasts produce approximately one non-collagenous molecule for each collagen molecule; however, the molecular weight of NCPs is on average around 50 kDa, whereas that of collagen in around 300 kDa, which explains the much greater proportion of collagen in the total mass. Beside their number, the fundamental role NCPs play in a number of complex biochemical reactions has recently become more evident, although this topic is still being heavily investigated (Heinegard and Oldberg, 1989; Behari, 2009).

Non-collagenous proteins are secreted by osteoblasts, during the various stages of their maturation. It has been proposed that pre-osteoblasts express *transforming growth factor beta* and *osteopontin*. Proliferating osteoblasts express proteins necessary to cell division, such as *histones, c-Fos*, and *c-Myc*, but also collagen type I. Mature osteoblasts express *alkaline phosphatase*. Differentiated osteoblasts express *osteopontin* and a closely related protein known as *bone sialoprotein* (BSP). During the mineralization phase osteoblasts express *osteocalcin*.

Transforming growth factor beta (TGF-β) is a protein that regulates cellular proliferation (Wang and Thampatty, 2006). In particular, TGF-β may induce apoptosis in various cell types. It has been observed that TGF-β activates both osteoclastic differentiation and osteoblastic activation (Chau *et al.*, 2009; McNamara, 2010). It should be noted that some of the so-called bone morphogenic proteins (BMP) belonging to the super-family of the TGF-β proteins, have the known effect of osteoblastic differentiation; the pathways of the BMP receptors and of the TGF-β receptors are coupled, and one modulates the other (Chau *et al.*, 2009). It has been suggested that the primary role of TGF-β in bone metabolism is indeed to ensure that osteoclastic and osteoblastic activations are properly coupled along the remodeling cycle (Matsuo and Irie, 2008). It seems that TGF-β also plays a role in deactivating the apoptosis mechanism in mature osteoblasts that differentiate into osteocytes (Noble, 2008).

Histones, c-Fos and c-Myc (proteins expressed by the FOS and MYC genes) play a key role in the differentiation of most cell types. Histones are the proteins that package the DNA, whereas c-Fos and c-Myc are transcription factors.

Tartar-resistant acid phosphatase (TRAP) and alkaline phosphatase (ALP) are two enzymes (proteins that catalyze a chemical reaction) whose precise physiological function in bone metabolism is unclear. Tartar-resistant acid phosphatase produces phosphate ester hydrolysis, and is found in significant quantities in active osteoclasts, in particular in their ruffled border. It is considered a valid marker for osteoclastic activity, as it has been observed that increased bone resorption is associated with increased released of TRAP from active osteoclasts (Watts, 1999). Alkaline phosphatase is a hydrolyzing enzyme responsible for a process called dephosphorylation, the removal of the phosphate groups from many types of molecules. Alkaline phosphatases are most effective in alkaline environments (pH $\approx$ 8). Diminished levels of ALP are associated with defective bone mineralization; however, the precise mechanism that this enzyme plays in the process is still unclear. What we know is that high serum levels of ALP may be an indicator of intense osteoblastic activity, since active osteoblasts secrete ALP in the extracellular space (Gordon *et al.*, 2007).

Osteopontin (OPN) is a protein synthesized by a large number of cells, including fibroblasts, osteoblasts, osteocytes, macrophages, skeletal muscle myoblasts, and a number of extraosseous cells in the inner ear, brain, kidney, and placenta (Heinegard and Oldberg, 1989). Osteopontin favors the adhesion of cells onto the surface where it is present (Standal *et al.*, 2004). This has been tested for various cell types, including rat osteoblast-like osteosarcoma cells, fibroblasts, and osteoclasts. Interestingly, its effect is inhibited by arginine-glycine-aspartic acid (RGD). It is believed that osteopontin anchors osteoclasts to the bone surface and develops their ruffled borders, which start the resorption process (Standal *et al.*, 2004). This protein seems to play several functions in bone physiology. Osteopontin binds to hydroxyapatite, inhibiting the growth of crystals (Standal *et al.*, 2004). It also appears to play a role in osteoclastic differentiation: while the deficiency of OPN does not reduce the osteoclasts count during bone formation, in pathological conditions, i.e. under the stimuli of parathyroid hormone (PTH), the lack of OPN prevents the rapid growth of the number of osteoclasts that is observed otherwise. The mechanism is not entirely clear, but OPN seems to interfere with the metabolic pathways of the receptor activator for nuclear factor κB ligand (RANKL) and with osteoprotegerin (OPG), two essential biochemical factors in the differentiation of macrophages into osteoclasts. Last, but not least, OPN plays a clear role in the modulation of osteoclast activity: OPN is found in high concentrations at the interface between old and newly formed bone, and also at the interface between bone and cells (*laminae limitantes*), where osteoclasts adhere (Standal *et al.*, 2004). Various studies suggest that OPN mediates this adhesion; where OPN is deficient, osteoclasts remain attached to the surface, and their mobility is drastically reduced.

The bone sialoprotein (BSP) is used as a marker for osteoblastic differentiation (Gordon *et al.*, 2007); its function in bone metabolism is still being investigated. Some evidence suggests that BSP plays a role in (i) osteoid mineralization, by providing a nucleation mechanism to hydroxyapatite, and (ii) osteoblastic differentiation (Hunter and Goldberg, 1993; Ganss *et al.*, 1999). A confusing result is that BSP gene knockout mice do have only minor osteoblast and mineralization defects (Gordon *et al.*, 2007); but this is a contradiction only in appearance. As in most physiological processes, there are always multiple mechanisms that act in parallel, and the lack of BSP is probably compensated for by an overproduction of other similar proteins. Some recent evidence suggests that the mechanism of BSP regulation might be more complex, as it requires close interaction between BSP, the osteoblast, and the extracellular matrix (Ganss *et al.*, 1999).

Osteocalcin is a low molecular weight protein, whose amino acid residuals provide strong calcium-binding properties. Thus, it is mostly found attached to hydroxyapatite. It appears to play a role in osteoid mineralization, by inhibiting the growth of hydroxyapatite crystals, and in osteoblast–osteoclast modulation, by contributing to the recruitment of osteoclasts (Kavukcuoglu *et al.*, 2009). However, none of these findings seems conclusive, and the true functions of osteocalcin are still a matter of investigation.

6.1.3   Inorganic matrix

The inorganic portion of the bone extracellular matrix is formed primarily by calcium phosphates (apatite), present as *hydroxyapatite* ($Ca_{10}(PO_4)_6(OH)_2$), a hydroxyl of the apatite group. It crystallizes in a hexagonal crystal system, where fluorides, chlorides, or carbonates can replace the OH^- groups. In bone, hydroxyapatite (HA) is sometimes present as calcium- and hydroxide-deficient apatite with many carbonate substitutions plus various other impurities. Bone HA is usually formed from relatively small crystals (≈ 200 Å), which makes it more prone to substitutions and to dissolution by osteoclasts. While geologic HA is a fairly stable substance, in bone because the crystals are so small most of the unit cells are on the surface, and this creates very favorable conditions for substitution by cations of the same size (K^+, Na^+, Mg^+, etc.) and by anions ($H_2PO_4^-$, HPO_4^{2-}, F^-, etc.), and make bone HA in general quite reactive. Evidence of such reactivity is the Ca:P ratio, which in bone HA is generally non-stoichiometric and varies considerably.

Hydroxyapatite binds to collagen at both molecular and ultrastructural levels. Thus, discussing the mechanical properties of bone HA alone is not particularly useful. It is more relevant to discuss the variation of tissue properties as a function of the degree of mineralization.

6.1.4   The mineralization process

When discussing mineralization, it is first important to define the context. Mineralization occurs during bone growth (*bone modeling*) and also in mature subjects (*bone remodeling*). When bone grows by endochondral ossification it is the cartilaginous tissue at the growth plate that mineralizes, whereas in mature bone it is the newly apposed osteoid that mineralizes. Here we shall focus on the latter process, osteoid mineralization.

Spontaneous precipitation happens only when a solution is oversaturated with mineral ions, but this is rarely the case in physiological conditions. Thus, the crystallization process must be initiated by some additional mechanism. In endochondral ossification, the HA crystals start to nucleate around small *matrix vesicles* that bud from the cytoplasmatic process of various cells. The presence of these vesicles has not been reported in osteoid mineralization, except in severely pathological conditions. Osteoid mineralizes first through the precipitation of tiny HA crystals in the gaps between collagen molecules in osteoid fibrils; what is driving this nucleation is unclear. Collagen alone does not seem to start the process. In addition, it has been observed that there is a delay between the apposition of the osteoid and the beginning of the mineralization process, which suggests that before HA can start to crystallize, some additional chemical reactions have to take place. The attention of researchers is primarily focused on the role that the other non-collagenous proteins play in this process.

Molecular and histochemical studies show that at the boundary between mineralized and unmineralized bone (the mineralization front) the most abundant proteins beside collagen are osteopontin, osteonectin, and bone sialoprotein (BSP) (Ogata, 2008). Other studies have also demonstrated the presence of phosphorylation and dephosphorylation enzymes. Of course, the higher concentration of a protein does not in itself automatically imply that this protein is the driving factor of mineralization. In-vitro studies showed that among these proteins BSP shows the most consistent nucleation capabilities (Gordon *et al.*, 2007; Hunter and Goldberg, 1993). It seems that BSP stabilizes the apatite nuclei, favoring nucleation, although in larger concentrations, it retards crystal growth. Other theories suggest that the lag period between osteoid apposition and the start of the mineralization process is because of the need to remove an inhibitor protein from the osteoid surface.

While the topic remains debated, it is clear that HA crystallizes over collagen fibrils, with its acicular crystals mostly oriented along the collagen fibril direction. The relationship between collagen type I and HA mineralization is intimate. In *osteogenesis imperfecta*, a genetic bone disorder where the collagen type I has quantitative and qualitative abnormalities, the mineralization process is also altered: crystals are smaller, their chemical composition is altered, and heterotropic ossifications are sometime observed.

However, collagen alone is not capable of starting the nucleation of HA crystals; indeed many other connective tissues made of collagen do not calcify. This opens up two possible scenarios:

- The conditions are sufficient for nucleation to start, but there are inhibitors that prevent the mineral deposition; the process starts only when these inhibiting substances are removed.
- There are other proteins that play a nucleation role; when present in appropriate concentration, they can begin the process.

In fact, we are probably dealing with a complex combination of these two mechanisms. Another non-collagenous protein, the matrix GLA protein (MGP), is expressed in high concentrations in cartilage and smooth muscle cells. MGP-knockout mice show massive calcifications in joints and blood vessels (Luo *et al.*, 1997). On the other hand, in addition to BSP, knockout studies also suggest that osteocalcin, type-X collagen, biglycan, and possibly other proteins may play a role. As usual, knockout studies also showed that it is nearly impossible to impair the mineralization process by knocking out a single gene. In the end, bone mineralization appears to be a complex, systemic, and highly redundant process, where HA, collagen, and many non-collagenous proteins are involved.

6.1.5   Matrix structure and organization

Being mineralized, the bone extracellular matrix cannot grow interstitially like cartilage. Everything that happens to bone must happen at its surface. In long bones, the endosteal surfaces can be distinguished as cancellous, endocortical, and intracortical (Haversian) surfaces. The largest surface fraction is taken up by the cancellous surface (60%), followed by Haversian (30%), and then by endocortical and periosteal (5% each) surfaces (Weiss, 1988; Cowin, 1988). Free surfaces are lined by cells, and traversed by blood vessels, nerves, and cell processes. As explained in Chapter 5, in most osteons the central cavity, called the *Haversian canal*, is occupied by blood vessels and nerves. Volkmann's canals, which connect Haversian canals to each other and with the periosteum, provide another channel for blood vessels. A third level of porosity is provided by the canaliculi, which usually host the cells' processes.

A first vital aspect, which has been fairly neglected until recently, is the strict relationship between each physiological process (growth, adaptation, repair) and the blood supply. All nutrients and most cellular progenitors are transported in the blood stream. In cancellous bone, capillary arcades are formed in the marrow space between trabeculae. In cortical bone, however, the space between free surfaces is too big, and thus the blood capillaries must penetrate the tissue through the Haversian canals. Although there is not yet quantitative evidence to sustain this assumption, I believe that the primary function of osteons is to ensure the necessary structural strengthening around Haversian canals, which are formed when the bone becomes too thick to allow effective nutrition only from the surface. All higher vertebrates have osteons, except very small animals like mice, where in many regions of the bone, the thickness of the cortex is so small that Haversian systems are not necessary, and thus no osteons form around them, and the compact bone is purely lamellar.

Bone innervations have also not been entirely explored. Nerve fibers are classified, among other criteria, according to the type of neurotransmitter they use. Inside Haversian canals and bone marrow, adrenergic nerve fibers (adrenaline neurotransmitter) have been reported; these are present in the secondary neurons of the sympathetic nervous system, which is responsible for various homeostatic regulation mechanisms. The discovery of β2-adrenergic receptors on osteoblasts (Takeda *et al.*, 2002), and that the receptor-mediated signaling in osteoblasts inhibits bone formation and triggers RANKL-mediated osteoclastogenesis and bone resorption, has opened a fundamental research front, which sees an integrative approach to the relation between energy storage in adipocytes, the sympathetic nervous system, and bone metabolism, through the mediation of *leptin*, a hormone secreted by adipocytes, which enters the central nervous system and binds with various neuron types. It has been speculated that this could be a systemic mechanism of bone mass regulation. This observation has also stimulated clinical research on the possibility of using beta-blockers to fight osteoporosis; observational studies produce controversial conclusions, although the majority of these studies seem to suggest a positive effect of these compounds on the reduction of fractures (Elefteriou, 2008; Reid, 2008).

If we study the histology of the bone extracellular matrix we can identify a portion of tissues that are encapsulated with a clear cement line, as if they were produced in a single biological process. In cortical bone, these units are the *osteons*, complete with their multiple lamellae with different collagen orientation, the Haversian canal, the various canaliculi, vascularizations, and innervations. In the cancellous bone, this structure is less evident, but we can identify a similar structure called a trabecular packet, or *hemi-osteon*, shaped like a shallow crescent of roughly the same radius of osteons, and with osteocyte layers between lamellae.

6.1.6 Fluid–matrix interaction

Bone can be described as a porous medium containing a solid phase, a fluid phase, and a cellular phase. The motion of the fluid phase within the pores is called interstitial fluid flow. As described in Section 6.3.3, early evidence of the direct mechanosensation of cells somehow forced the assumption that considerable drag deformations had to be induced on the bone cells by the interstitial flow; however, only recently has experimental evidence on this flow and on its interaction with the bone extracellular matrix been obtained.

Firstly, there is the wide range of characteristic diameters that the interstitial fluid network shows. We go from the 250 micrometers of Haversian canals, to the 50 micrometers of a Volkmann's canal, to the 1 micron of lacunae, to the 0.05–0.10 micrometers of space between the canaliculi wall and the surface of the osteocyte process. To ensure a sustainable lifecycle, every cell in the bone tissue must be reached by nutrients and oxygen, and its metabolites should not be allowed to accumulate around the cell. As bone is permeated by living cells not only on its free surfaces, but also inside its matrix (osteocytes) this means that interstitial fluids must flow over the entire aforementioned network. But how can this happen, in the absence of a true circulatory system, such as that that ensures the circulation of blood?

There are essentially three possible circulation mechanisms for interstitial fluid: the first can be called "passive" and comprises diffusion (Brownian motion), concentration gradients, and convective transport. This passive transport mechanism can be represented with diffusion–convection equations, which account for diffusion, convection, uptake, and reaction. The second mechanism is called "intracellular" and relies on the fact that the majority of the cells that permeate the bone extracellular matrix are connected via cytoskeletal processes, like dendrites connected by gap junctions. The third mechanism is called "active" and is caused by the "pumping" effect that the cyclic deformation of the bone matrix during repetitive tasks, such as walking, induces on the interstitial fluid that permeates the matrix porosities (Lemaire *et al.*, 2006).

Many authors have questioned the importance of passive interstitial fluid circulation in bone cell metabolism. The problem is that the journey that a molecule of nutrient must take from the blood stream to an osteocyte cell encased deeply in the mineralized matrix seems tortuous, at least; while very small molecules may eventually complete the trip, large molecules would have a hard time reaching the osteocyte lacunae. The nutrient must first permeate out of the blood capillary that penetrates the Haversian canal. These capillaries have a continuous epithelium lined with basement membrane, a thin sheet of collagen fibers; the porosity of this membrane is unknown, but the epithelial fenestrae (small pores in the epithelial cells) in bone marrow capillaries are around 10 micrometers, and the porosity of the basement membrane of the capillaries in the intestinal region is 2–9 micrometers (Takeuchi and Gonda, 2004). Lining cells usually cover the free surfaces of the mineralized matrix; if this is also true for the Haversian canal, then the nutrient molecule also has to pass through this barrier to enter the interstitial space. But the biggest difficulty is probably the passage through the canaliculi, which are partially occupied by the osteocytarian processes; it has been estimated that the free space between the canaliculi wall and the process surface is between 0.05 and 0.1 micrometers. Thus it is difficult to imagine that anything larger than 100 nm could reach the osteocyte via the interstitial space.

In theory, another pathway could be intracellular: the nutrients could penetrate the membranes of the lining cells, and then be transported to the osteocyte via the osteocytarian processes that link them to the surface. However, the intra-membrane space of the processes is also 80–100 nm, and in addition the inter-process gap junction size has been theorized to be around 2 nm (the mechanism of permeability regulation at osteocyte gap junctions is poorly understood, and might be much more complex, owing to the mediation of gap junction proteins, such as Connexin 43). Thus, while some intracellular transport is clearly possible, interstitial transport remains the prime candidate.

A somewhat related mechanism has been proposed, where molecules are transported in the extracellular space, but actively by osteocytes. It has been observed that the surface of the osteocyte processes is covered by micro-filaments that show some contractile properties; thus we can imagine that these fibers contract progressively along the length of the process, pushing small molecules through the canaliculi (Knothe Tate, 2003).

All these mechanisms appear insufficient to explain the amount of bone interstitial fluid flow: experiments with radioactive tracers have suggested that the fluid clears the canaliculi over 2000 times in 24 hours (Rowland, 1966). This is also why much research is focused on active transport. While we move, our bones deform, and the extracellular matrix accelerates the interstitial fluid that permeates it through a solid–fluid interaction. Owing to the difficulties of measuring the flow of a fluid inside a hard tissue with porosities of 1 micron, most studies are based on mathematical models, or on indirect measurements, such as strain-induced streaming potential. But with all their limitations these studies confirm that during normal physical activities a considerable amount of fluid is pumped in and out of the interstitial space, down to the canalicular network (Lemaire *et al.*, 2006).

6.2   Bone cells

Bone cells are the subject of intense research, with dozen of papers being published every day. Trying to provide an up-to-date bibliography would be a futile exercise. However, most of the information provided in this section is well-established and can be found in many textbooks. For further reading, recommend among others, the monumental book edited by John Bilezikian (Bilezikian *et al.*, 2008).

A variety of specialized cells participate in the growth, maintenance, and adaptation of bones. Here I shall focus only on three cell types, which play a key role in the production and repair of the bone extracellular matrix: *osteoclasts*, specialized cells that can resorb the mineralized matrix; *osteoblasts*, specialized cells that synthesize collagen type I and many other proteins that constitute the osteoid, and drive its mineralization; and *osteocytes*, specialized cells that permeate the bone extracellular matrix, forming a network that plays a vital role in a number of mechanobiological processes. Some authors also consider a fourth type of cell, called lining cells, which are quiescent osteoblasts, and cover the extracellular matrix surface with no remodeling cycles active in that region; however, owing to their strong similarity with osteoblasts and osteocytes, I shall not discuss them separately.

Although osteoblasts and osteoclasts appear to be opposite cells, with competing functions (one makes, the other destroys) they have many common features. They are both highly specialized cells, with a relative short life span. Both are produced by line differentiation from progenitor cells *in situ*, at the locality where they are needed; thus their mobility is quite limited, and unlike their progenitors, they do not generally enter the blood stream. Indeed, all remodeling processes start with the recruitment *in situ* of progenitor cells, and their differentiation.

6.2.1 Osteoclasts

6.2.1.1 Cell line

The proliferation of osteoclasts (Figure 6.1) has never been documented. Osteoclasts derive from the mononuclear–phagocytic lineage of the hematopoietic marrow, the granulo-monocyte colony-forming unit (GM-CFU). Isolated hematopoietic cells proliferate, differentiating into GM colony-forming units, which in turn proliferate, differentiating into pre-osteoclast mononuclear cells. These pre-osteoclasts fuse into a single quiescent osteoclast, which polarizes on contact with the matrix surface (i.e. the apical and basal portions of the cell membrane become different and take on specialized functions) and becomes active.

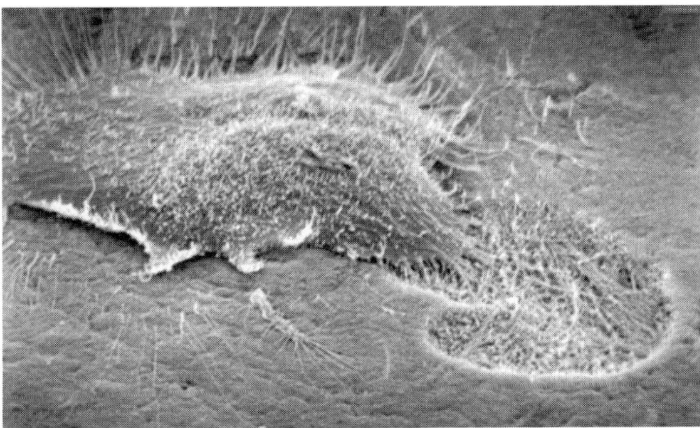

Figure 6.1   Scanning electron micrograph of a rat osteoclast with adjacent resorption pit; image width, 75 micrometers. Courtesy of Professor Tim Arnett, UCL.

6.2.1.2　Morphology and function

An osteoclast is a large, multinucleated cell, with up to 50 nuclei. Its cytoplasm appears like spume, owing to the large number of vacuoles and vesicles it contains. In size it ranges between 20 and over 100 micrometers.

The osteoclast's primary function is to resorb bone tissue. The peculiar morphological traits of osteoclasts become evident only when they are active, which happens when they attach to the mineralized matrix. Active osteoclasts are found in small cavities, called Howship's lacunae, formed inside mineralized matrix that they resorbed themselves. Near the matrix surface, the osteoclast becomes striated and the border gets ruffled, owing to considerable membrane enfolding. The ruffled border is surrounded by ectoplasm (the outer layer of the cell cytoplasm) with no organelles but many active filaments, called the *clear zone*. Osteoclasts adhere to the matrix at the clear zone with a specialized structure called a *podosome*, sealing a portion of the matrix surface into a chamber, inside which the ruffled border secretes chemicals capable of attacking bone. Its attachment is mediated by *integrin*, *vitronectin*, and collagen receptors.

Once the podosome is sealed, the ruffled membrane is penetrated by many microtubules that will deliver vesicles into the resorption space. Acidification occurs, from the release of carbonic anhydrase II, an enzyme that catalyzes the rapid conversion of carbon dioxide and water to bicarbonate and protons, and by proton pumps that transfer the protons from the resorption space to inside the osteoclast membrane (from where they are then expelled in the extracellular space via ion channels on the basal (opposite to the matrix) membrane, so as to maintain cellular electroneutrality). The severe acidification of the resorption space dissolves the HA, exposing the collagen matrix to the attack of various matrix-degrading enzymes released by the osteoclast, including *cathepsins* and *collagenases*. Once dissolved, the inorganic matrix and the degraded matrix proteins are absorbed into transcytotic vesicles, which transport them through the osteoclast to the basal membrane from where they are expelled into the extracellular space.

6.2.1.3　Regulation

This differentiation process requires a certain bone microenvironment, and in particular the concurrence of osteoblastic differentiation. Pre-osteoblastic–stromal cells secrete macrophage colony-stimulating factor (M-CSF), which favors the proliferation of hematopoietic cells into GM-CFU precursors. The second mechanism was already mentioned in the description of osteopontin protein. Osteoblasts, bone marrow stromal cells, and activated T cells secrete the receptor activator for nuclear factor κB ligand (RANKL), which binds to the RANK receptor present on osteoclastic precursor cells. This bond creates a cascade of events essential for the differentiation of osteoclastic precursor cells and their fusion into multinucleated cells, as well as for the activation and survival of mature osteoclastic cells.

The same mechanism also exposes one of the most interesting regulation mechanisms for osteoclasts. The same cells that produce RANKL can also produce *osteoprotegerin* (OPG), a glycoprotein cytocine (a signaling molecule secreted by cells mostly for communication purposes). Osteoprotegerin is a decoy receptor for RANKL; when OPG is released RANKL binds to OPG instead of the true RANK receptor, preventing the RANKL activation of osteoclasts and their precursors. The modulation of OPG and RANKL enables fine-tuning of the rate of differentiation, activation, and apoptosis of osteoclasts; in a way osteoblasts, and their precursors, control osteoclastogenesis and osteoclast activity, which creates a very interesting interdependency between these two cell lines.

In addition, a number of other cytokines and hormones interfere with this mechanism, up- or down-regulating M-CSF, RANKL, and OPG: TGF-β and interleukin-1 stimulate M-CSF production (so increasing the number of pre-osteoclastic cells) and increase their expression of RANKL. In addition, TGF-β increases OPG production, parathyroid hormone (PTH) increases RANKL production and decreases OPG, 1,25-dihydroxyvitamin D_3 increases RANKL production, and estrogens increase OPG production.

6.2.2 Osteoblasts

6.2.2.1 Cell line

The differentiation process that produces osteoblasts (Figure 6.2) is more difficult to investigate. Firstly, osteoblasts are active in the embryo, during skeletal growth, and in adults during bone remodeling and bone healing. Secondly, osteoblasts derive from a line of mesenchymal stem cells that differentiate into a variety of collagen-secreting cells, including fibroblasts and chondroblasts. We can positively identify an osteoblast precursor only when it develops bone-forming properties, which happens relatively late in the differentiation.

The cell line begins with mesenchymal stem cells, multipotential mesenchymal precursors. To be precise, this name refers to the mesenchyme, a reticular connective tissue present only in the embryo. In mature individuals, the source of stem cells is

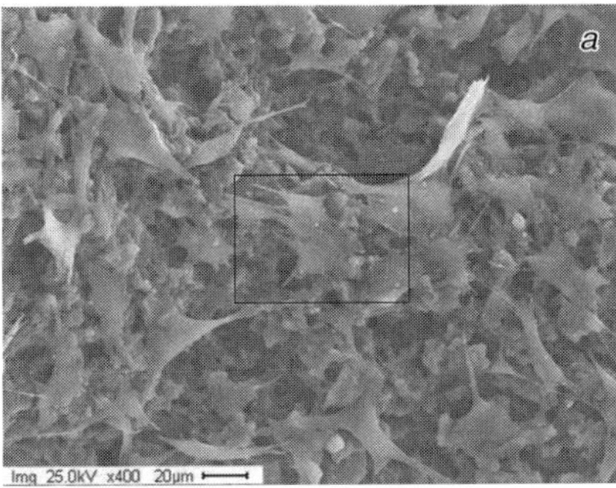
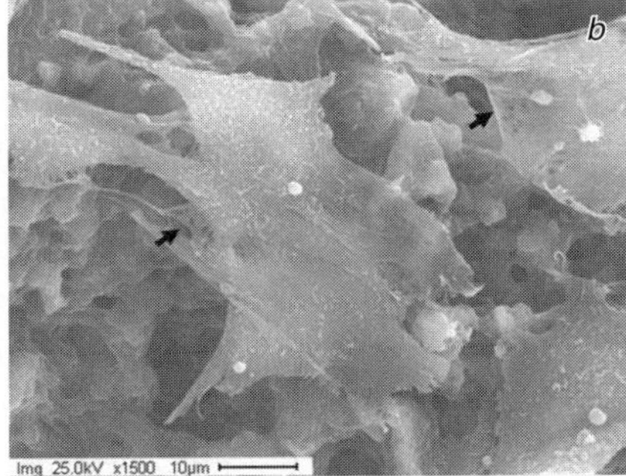

Figure 6.2 (a) Osteoblasts proliferating on PCL-HA scaffolds (400×). (b) Enlargement of panel (a) showing the details of the attached osteoblasts with a meshwork of extracellular matrix (arrows) and cellular projections (1500×). Reprinted from (Yu *et al.*, 2009) with permission granted by the Creative Commons Attribution License, © 2009 Yu *et al.*; licensee, BioMed Central Ltd.

usually (but not exclusively) the bone marrow, which gives rises to both hematopoietic line such as osteoclasts, and non-hematopoietic (stromal) cells, including osteoblasts. Thus, in this case it is probably more accurate to refer to these progenitors as *multipotent stromal cells.* The multipotent cells differentiate into an immature cell called an *osteoprogenitor,* which is recognizable because it expresses a regulatory transcription factor Cbfa1/Runx2 (transcription factors are proteins that bind to DNA and regulate gene transcription into mRNA, and thus into proteins). Osteoprogenitors start to proliferate and differentiate into osteoblasts. Osteoblasts are marked by their expression of bone-specific proteins, such as bone scialoprotein or osteocalcin.

For many years, osteocytes were considered a third type of cell, owing to their peculiar location, morphology, and function. However, today we know that osteocytes are formed by a further differentiation of mature osteoblasts, undertaken once trapped inside the extracellular matrix. Still, in terms of function, it makes sense to discuss this cell type separately in a dedicated section.

6.2.2.2   Morphology and function

Osteoblasts are small, rounded cells, with a cytoplasm rich in organelles. In the active state they have a cuboidal shape, rarely undergo mitosis, and have a large nucleus, cellular processes, gap junctions, abundant endoplasmic reticulum, an enlarged Golgi complex, and many secretory vesicles containing protocollagen. Osteoblasts express the vast majority of bone proteins, including collagen type I, alkaline phosphatase, osteocalcin, osteopontin, and osteonectin.

6.2.2.3   Regulation

The proliferation and differentiation of osteoblasts is regulated by a number of growth factors and transcription factors.

The primary growth factors that regulate osteoblasts are TGF-β, *bone morphogenic proteins* (BMP), belonging to the super-family of TGF-β proteins, and *insulin-like growth factors* (IGFs). The problem is that none of these growth factors is specific to osteoblasts, and they seems to play a key role primarily in coupling the osteoblastic regulation with that of other cell types, namely osteoclasts and osteocytes. Thus, their role is usually played in combination with other hormones, such as parathyroid hormone, prostaglandin E2, glucocorticoids, sex steroids, etc.

Various authors have also found that some proteins related to angiogenesis were also involved with osteoblast differentiation: among others, vascular endothelial growth factor (VEGF), platelet-derived growth factor (PDGF), and fibroblast growth factor (FGF) are all secreted by blood vessels in the proximity. Vascular endothelial growth factor promotes osteoblastic differentiation, and speeds up mineralization; PDGF seems to be present only in association with inflammation and repair, during which it also has a significant mitotic effect; while FGF stimulates osteoblast replication, but tends to inhibit collagen synthesis. It is not at first obvious why proteins related to angiogenesis should play a role in bone; this will become clearer in the following section, which describes bone mechanobiology.

Transcription factors also play a key role, through the so-called *osteogenic signaling pathway*. The first two are Runx2 and Osterix. Both appear essential for the differentiation of osteoclasts. But this is only the tip of the iceberg: these two factors are themselves regulated by a number of other factors, and there are a number of pathways that have recently been identified and are now under close scrutiny. These include the BMP-Smad, the MAPK, the AKT, the Wnt-β-catenin, the notch, and the leptin pathways. A detailed discussion of these pathways is beyond the scope of this book; some mechanisms will be further discussed in Section 6.3.

6.2.3 Osteocytes

6.2.3.1 Cell line

Osteocytes (Figure 6.3) are terminally differentiated osteoblasts trapped in their secreted matrix.

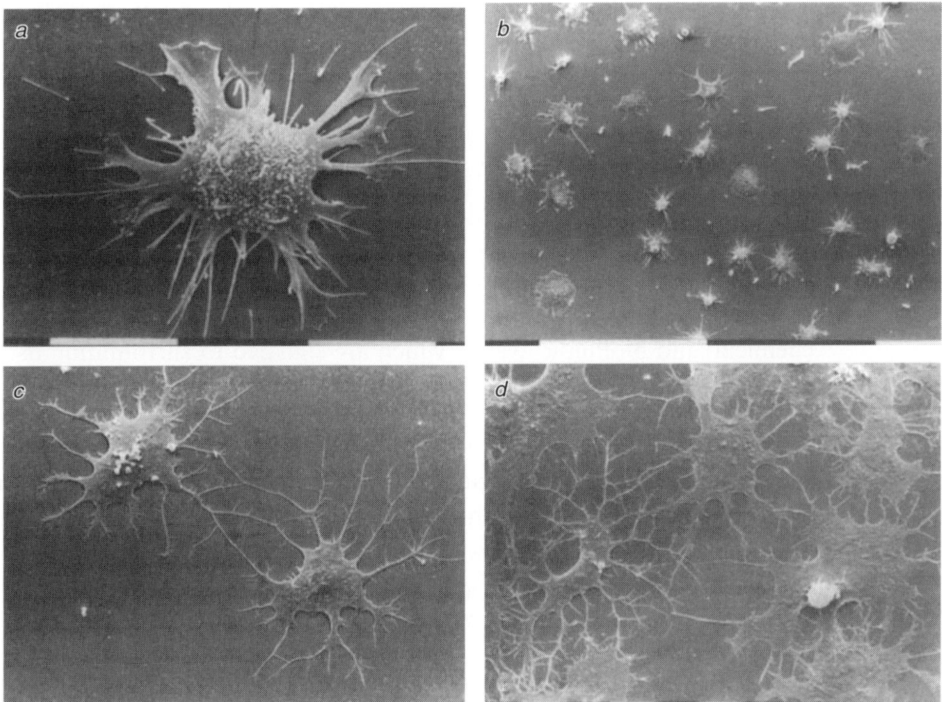

Figure 6.3 High and low magnification photomicrographs of osteocytes in culture. (a) and (b) Osteocytes after two hours in culture (bar = 100 micrometers). (c) Two osteocytes after one day of culture, appearing to make contact with each other (bar = 10 micrometers). (d) Extensive network of flattened osteocytes with many branched cell processes after two days (bar = 10 micrometers). Reprinted with permission (van der Plas and Nijweide, 1992).

6.2.3.2   Morphology and function

Osteocytes are small rounded cells with long cytoplasmatic processes that penetrate deeply into the canaliculi connecting the osteocyte to other surrounding osteocytes and to the quiescent osteoblasts (lining cells) on the bone surface. Processes are connected to each other by true gap junctions, similar to those between neuronal cells.

While the size of the lacunae seems quite similar in vertebrates (5–20 micrometers in mice, 9–20 micrometers in human beings) the density of osteocytes can be quite different from species to species (31 900 mm^{-3} in bovine bone to 93 200 mm^{-3} in rat). The literature provides contradictory results, mostly in terms of areal density, which can be misleading, as the spatial distribution of osteocytes can be quite anisotropic. Human osteocytes seem to have a volumetric density of 15 000 mm^{-3}, which is less than the values reported for animals. Since it is hard to explain the dramatic differences between species (the osteocyte density should be mostly defined by the transport mechanisms of nutrients and signaling factors, which depend on lacunar and canalicular dimensions that are more or less constant across species), we suggest there might be some serious methodological limitation in the way we currently measure this density.

Osteocytes are incapable of mitotic division, and the ability to secrete matrix proteins is somehow lost, but they are quite surprisingly capable of small, rapid resorptive activity, much like osteoclasts, although they do not use any of the osteoclastic specialized organs; this is called *osteocytic osteolysis*. This mechanism is probably necessary for the formation of the osteocyte lacunae, and to prevent the lacuna space from being invaded by HA crystals during osteoid mineralization.

We are now convinced that the primary role of osteocytes is signaling. They permeate the whole bone matrix in large numbers (some authors estimate that there are ten times more osteocytes than osteoblasts) and behave like a network of sensors that are capable of monitoring the entire bone matrix for biomechanical, biochemical, and biofluidic events, which are signaled to surrounding osteocytes and up to the bone surface at the lining cells, where eventual remodeling processes are initiated.

6.2.3.3   Regulation

The regulation of osteocytes remains riddled with questions. Osteocytes are low-metabolism cells that do not replicate; some authors suggest that they have a half-life of 25 years. In addition, all we know about osteocyte regulation comes from experiments on cell cultures; but owing to the difficulty of isolating true human osteocytes (which are normally trapped inside the mineralized matrix), most of the work is done on immortalized cell lines, like the MLO-Y4 of murine origin, and so any result must be treated with caution.

Some regulatory mechanisms seem related to the ability of osteocytes to keep the canaliculi and the lacunae viable. For example, the matrix metalloproteinase-2 (MMP-2) seems quite important for the formation of the osteocyte network. Matrix metalloproteinase acts like collagenase, breaking up various collagen types including type I. In MMP-2 knockout mice there is a drastic reduction of the number of canaliculi.

Osteocytes show some specificity for the dentin matrix protein-1 (DMP-1); in knockout studies a connection was found between this protein and fibroblast growth factor 23 (FGF-23), which is known to regulate phosphate absorption by the kidney, suggesting that osteocytes can have a role in mineralization control and phosphate equilibrium, remotely controlling kidney function.

Under mechanical stimuli, osteocytes were found to produce matrix extracellular phosphoglycoprotein (MEPE) quite late after the stimuli, which is suspected to inhibit bone formation.

The lack of expression of the klotho enzyme encoded in the KL gene produces accelerated aging. The phenotype of this aging model is characterized by many alterations of the osteocytes. We do not know the precise mechanism yet, but the observation that the klotho enzyme has a co-receptor function for FGF-23 opens some interesting hypotheses.

Osteocytes seem also to be regulated by oxygen. Osteocytes have higher concentrations than osteoblasts of oxygen regulated protein, ORP-150. This aspect will be further discussed in Section 6.3.

Osteocytes have receptors for estrogens, parathyroid hormone, vitamin D_3, corticosteroids, and TGF-β. Interestingly, so far no one has reported receptors for insulin, insulin-like growth factors, or growth hormones.

Osteocytes have some traits in common with elements of the nervous system. For example, the presence of glutamate transporters was found inside osteocytes, and these produced nerve growth factor in the presence of a bone fracture.

6.3   Bone mechanobiology

6.3.1   Introduction

The long description of the bone cells and extracellular matrix should not confuse our primary scope, that of building a multiscale model of the skeleton. In this context, what we are interested in at the tissue-cell level is what connects the biomechanical function of the skeleton to the skeletal adaptation and repair mechanisms we can study in great detail at the tissue-cell scale. This, of course, takes us immediately to the interface between the mechanical stimuli, and the biological answer, i.e. mechanobiology.

A specific note on terminology: hereinafter I shall use the term *bone modeling* to indicate organogenesis, i.e. embryonic formation and growth in immature individuals. The term *bone remodeling*, on the contrary, will be used to indicate the metabolic processes that constantly destroy and regenerate the bone extracellular matrix in mature individuals. This process can produce no variation of the total skeletal mass (homeostasis) or variations of the total skeletal mass. Lastly, I shall use the term *bone adaptation* only when I want to suggest that a bone remodeling process that produced variations of the total skeletal mass was due to the changes in the environmental conditions, thus implying that the changes in bone mass are an adaptation to such environmental variations.

6.3.2 Physiology of bone remodeling

The bone remodeling cycle (Bilezikian *et al.*, 2008) starts at a certain location of the extracellular matrix surface during the *activation* phase. A portion of the matrix is destroyed during the *resorption* phase. The resorption cycle is terminated by a *reversal* phase, followed by a *formation* phase.

6.3.2.1 Activation

For a long while, the attention of researchers was focused mostly on resorption and formation phases. The activation phase was described in most biology books in a way that I always found unsatisfactory. All we knew was that bone matrix was covered by a layer of resting osteoblasts (lining cells) attached to a thin layer (0.1–0.5 micrometers) of unmineralized, collagen-poor connective tissue called the endosteal membrane. At activation, the lining cells digested this membrane, and detached from the matrix surface, opening the way to osteoclasts, and then to active osteoblasts. How, these cells managed to get there was a mystery.

Permit me a small personal digression. In the late 1980s, when I began my research career, my first research topic was bone remodeling. With the enthusiasm and naïveté appropriate for my young age, I started to read everything on the subject of bone remodeling that I could put my hands on. One point in particular puzzled me, a young mechanical engineer converted to biomechanics. All authors agreed that the primary raison d'être for bone remodeling was maintenance of the bone matrix. Of course, this implies that during our daily life the bone matrix is constantly damaged, and thus requires some repair. Indeed, some authors began to report the presence of many micro-cracks in the bone matrix of normal bones. So bone remodeling was all about repairing micro-fractures of the bone matrix. But if this was the case, why were bone remodeling and bone fracture healing described as two completely different processes? In particular, what I found peculiar was that in every fracture healing model the first step was an inflammatory response that hyper-vascularized the region around the fracture. On the contrary, in bone remodeling models the word "blood" never appeared. Apparently, by some sort of magic, osteoclasts and osteoblasts appeared at the site of bone remodeling and started to perform their functions. The concept of a bone multicellular unit, first introduced by Harold M. Frost, somehow implied a physical compartment, which however no one described. The communication between cells remained confined, but it was not clear how this could happen, given the idea that everything happened in an open extracellular space above the matrix. After these early years, I turned my attention to other research topics, but bone remodeling always remained my first love; with less persistence I managed to keep an eye on the relevant literature, but my problem with the activation phase remained.

Imagine my happiness when Ellen M. Hauge and colleagues at the Department of Pathology of Aarhus University described a sinus, which they called a *bone remodeling compartment* (BRC), that apparently wrapped sites at the matrix surface where active bone remodeling was taking place (Hauge *et al.*, 2001). As often happens, soon afterwards Parfitt (2001) reported in an editorial that this canopy of cells over the bone remodeling site was first described by Rasmussen and Bordier in their 1974 book *The Physiological and Cellular Basis of Metabolic Bone Disease*. But Hauge *et al.* did something more: they showed how this bone remodeling compartment was a conduit for osteoclast precursors. Parfitt suggested that this was a case of neoangiogenesis, implying that the BRC was connected to a newly formed blood supply, through which all cellular precursors and many systemic biochemicals necessary to the bone remodeling cycle reached the region of matrix to be remodeled.

Today, thanks also to the recent paper from Andersen *et al.*, we can confirm that bone remodeling activation starts with the lining cells covering the portion of matrix surface to be remodeled lifting from their attachment, to form a closed sac around the site, which is then connected with capillaries (Andersen *et al.*, 2009). This BRC is formed of tightly packed cells, which are positive to osteoblast markers, and remain connected at the periphery of the BRC to the layer of lining cells, actually sealing the BRC from the marrow space. The formation of capillaries explains the relationship found between bone cells and angiogenesis factors; bone-remodeling activation also involves angiogenetic mechanisms. The same group also provided a negative confirmation, observing that in multiple myeloma the BRC canopy is frequently disrupted; in these cases we find bone lesions that are not repaired (Andersen *et al.*, 2010). Thus, the formation of the BRC is necessary to bone remodeling.

6.3.2.2   Resorption

The BRC creates a conduit between the capillary system and the bone matrix surface, which is now fully exposed. Osteoclast and osteoblast precursors reach the BRC space and it is only through an intense signaling between these cells, the resting osteoblasts forming the BRC canopy, and other cell types present in the BRC, such as macrophages, that the resorption process starts.

Describing these signaling mechanisms as an ordered sequence makes their presentation clearer, but might be misleading: we should not imagine that this is an ordered sequence of events. I believe that it is more convenient, and probably closer to the truth, to imagine that at any point in time within the BRC all cell types are present, precursors or not. The dynamics of the regulatory mechanisms increase or decrease their number and their level of activity during the various phases of the remodeling process.

The BRC forms in an anatomical space that is usually filled with bone marrow. Thus it is reasonable to assume that the region will have a discrete number of stromal-line cells, including pre-osteoblasts. These secrete macrophage colony-stimulating factor, which stimulates the hematopoietic line of marrow cells as well as those transported in the blood flow, with the creation of granulo-monocyte colony-forming units that differentiate into mononuclear pre-osteoblasts.

Now it is time for the RANKL-RANK-OPG regulation to kick-in (Khosla, 2001; Theoleyre *et al.*, 2004). RANKL is expressed mostly by quiescent osteoblasts, i.e. lining cells. When it binds to the RANK receptor of the pre-osteoclasts, it activates the pathway that forces them to fuse into the multinucleated osteoclast.

It is probably pure chemotaxis that attracts the newly formed osteoclasts to the bone surface, where they polarize and start to resorb bone matrix. Osteoclasts tend to cut a cylindrical cavity into the bone matrix, called Howship's lacunae in cancellous bone, and a cutting cone into cortical bone. The cutting cone elongates with a typical speed of resorption of 20–40 micron day^{-1}, and expands radially with a speed of 5–10 micron day^{-1}. The resorption cycle typically lasts 1–3 weeks, leaving cutting cones 100 micrometers deep in cortical bone, and 60 micrometers deep in cancellous bone (van Oers *et al.*, 2008).

6.3.2.3   Reversal

Between the end of the resorption and the beginning of the formation there is a period of 1–2 weeks during which the cutting cone shows no osteoclasts, but various mononuclear cells of unclear origin. Some authors claim that these are macrophages, whereas others suggest they might be quiescent osteoblasts. This is a minor detail, as some of the cell-to-cell signaling between osteoclasts and osteoblasts requires cell contact, which would not be possible inside the cutting cone if all osteoclasts were to disappear from the cavity before the first osteoblast appears. I remain convinced that the best model is to assume that all BRC, including the cutting cone connected to it, contain virtually all cell types, although with various concentrations depending on the various phases of the remodeling process.

It is reasonable to presume that during the reversal period, osteoclasts and osteoblasts interact through some sort of signaling (Papachroni *et al.*, 2009; Matsuo and Irie, 2008). For example, stromal-line cells, including active osteoblasts, can also express OPG, which bind RANKL, preventing the RANK pathway from being activated, inhibiting osteoclast differentiation and activation. But the osteoblast–osteoclast signaling is much more complicated. In the transition phase alone there are dozens of signaling candidates, including matrix-receptor communication for TGF-β, BMP, IGF-II; cathepsin K; TRAP ↔ GPC4 homolog; TRAP ↔ TRIP-1; Atp6v0d2; sphingosine 1-phosphate ↔ S1P receptor; Wnt ↔ LPR5; ephrinB2 ↔ EphB4; connexin ↔ connexin; etc. All this messaging can be passed between cells by direct contact, via gap junctions, or by means of diffusible paracrine factors. In addition, we should not neglect the fact that the by-products of osteoclastic resorption contain many non-collagenous proteins that we have seen can play significant roles in bone-cell regulation.

Thus, even if we do not know exactly how osteoblasts and osteoclasts know each other's status, there are plenty of possible mechanisms to justify the reversal phase.

6.3.2.4 Formation

As the osteoclasts retract from the bone surface, and the macrophages remove the by-products of their resorptive activity, the resorption space is invaded by active osteoblasts, which start to secrete collagen and other proteins to form new osteoid. This phase is much slower, and it may take two or three months before the resorption space is filled up with new osteoid. As soon as it is formed, osteoid undergoes a rapid transformation, which takes it to approximately 70% of its mineralization in one to two weeks; the remaining 30% happens in a much longer time frame, typically three to six months.

6.3.3 Modeling mechanobiological processes

The skeleton undergoes a continuous cycle of resorption and regeneration even after growth terminates. It has been estimated that 1–3% of skeletal mass is destroyed and regenerated every year (Shea and Miller, 2005). Other authors, however, suggest bigger fractions, up to 10% per year (Watts, 1999). Everybody agrees on the great variability, and that rates might vary considerably in certain regions of the skeleton. Being a process that occurs at free surfaces, the higher the porosity, the higher the remodeling rate: although cortical bone provides 80% of skeletal mass, up to 25% of all cancellous bone is remodeled every year, compared with only 3% for cortical bone (Watts, 1999). Normally this process yields a net balance (or more accurately a small negative balance) of the bone mass, but under altered external conditions the skeleton is capable of adapting, increasing or decreasing the skeletal mass considerably.

In particular, it has been observed that there are mechanisms that regulate the shape, density, and mass of specific skeletal regions, owing to changes in the loading that the skeleton experiences daily. This bone adaptation to the biomechanical environment has fascinated generations of researchers and remains, in my opinion, one of the most exciting research topics a young bioengineer can embrace.

What probably makes this argument so interesting is that it is very complicated. Still today, in spite of the huge number of observations, elaborations, experiments, and theories produced, bone remodeling remains an elusive target. Thus, it is probably fair to separate in this book what we know from what we believe.

6.3.3.1 Effects of underloading

A very large number of experiments have been conducted on both animal and human models to understand if and how the reduction of mechanical loading reduces the skeletal mass (some systematic reviews can be found in (Bikle *et al.*, 2003; Giangregorio and Blimkie, 2002; LeBlanc *et al.*, 2007; Sievanen, 2010; Skerry, 2008)). In animal models, load reduction is induced by hind limbs unloading with tail suspension, by casting or other means of immobilization, by selective resections of tendons or nerves, and by surgical procedures that completely isolate a segment of skeleton from mechanical loading, while retaining all biochemical and biological connections. In human beings, it is quite difficult to create controlled conditions in healthy subjects within the limits of modern ethics: most data come from bed-rest experiments, and from subjects who spend time in space under weightlessness conditions.

This huge body of observations tends to draw a fairly consistent picture. If the load acting on the skeleton is drastically reduced, bone mass decreases over time with a negative asymptotical trend toward a lower limit. The resorption rate varies as a function of the animal type and the load reduction method, and it is not homogeneous across the skeleton: each bone shows a different resorption rate, and some suggest that the more load-bearing is the bone, the more sensitive it will be to load reduction. But in spite of all these differences there is always a lower limit to the bone mass that is never crossed, no matter how severe is the load reduction or how long it lasts in time.

Larger reductions of bone mass are primarily observed when there are some concurrent pathological conditions, such as in osteoporosis.

6.3.3.2   Effects of overloading

Overloading experiments are much more complicated, and thus less numerous (for reviews see the *Proceedings of the International Symposium on Physical Loading, Exercise, and Bone*, published in a special issue of *Bone* (Kannus *et al.*, 1996); also (Burr *et al.*, 2002; Qin *et al.*, 2010)). A first group includes those experiments involving strenuous exercise. The animal, or the human volunteer, is subject to a strenuous program of physical activities, much more intense than their normal lifestyles involve. These experiments can only be seen as qualitative, or at most semi-quantitative, since the relationship between the activity program, and the change in the loading that the skeleton actually experiences, is unknown. We control the activity program, which we know is related to the intensity of the loading, but we do not know this relationship, so in reality we do not have control. However, these experiments are useful for showing, in a totally undisturbed organism, that there is some positive correlation between increase in physical activity and increase in bone mass.

The second type of experiments involves a direct application of force to a segment of the skeleton, and quantification of how its bone mass adapts to changes in the loading regime. Of course, to apply force to a bone we need to be invasive, and in doing this we alter the local biological and biochemical milieu of the bone; this can only partially be compensated for by using controls subjected to an identical sham preparation, since we cannot exclude the effect of surgical invasion, or that of increased loading. However, these experiments provide us with true quantitative relationships between duration, intensity, direction, and frequency of loading and the bone adaptation that these experiments produce. While most experiments find a positive correlation between an increase in load intensity and frequency and the increase in bone mass, and some even confirm that this additional bone mass is added spatially to strengthen the bone to that loading condition (which indicates a true adaptation), overall attempts to quantify this relationship have produced contradictory results. In addition the difficulty of the experiments makes it very difficult to exclude the possibility that some measurements are not biased; we cannot rule out the possibility that some of the results provided in the literature are plainly wrong, which of course confuses the scenario.

6.3.3.3 Modeling bone adaptation at the organ scale

Although it was first Galileo Galilei (1564–1642) who in his *Discorsi e dimostrazioni matematiche intorno a due nuove scienze* suggested the relationship between the size of bones and the loads they had to sustain, the roots of the early theories of bone adaptation are in the scientific revolution of the eighteenth century. In 1809 Jean Baptiste Lamarck, in his *Philosophie Zoologique*, proposed that evolution occurred as organisms adapted to the environment, and then passed these adaptation traits to their offspring. While the Lamarckism concept of evolution by adaptation was soon after replaced by the much stronger Darwinian concept of evolution by natural selection, the idea that the form that life assumed a manifestation of its functions became a landmark idea for a good part of the nineteenth and twentieth centuries. Probably the most important book on this conception is D'Arcy Thompson's *On Growth and Form* of 1917.

Between 1830 and 1842 Auguste Comte wrote the *Course in Positive Philosophy*. While the recognition that the scientific method is superior to any other metaphysical approach in the investigation of nature seems self-evident to any modern scientist, some radical interpretations of *positivism* produced very negative effects in the science of the twentieth century, which still appear in the thinking of many researchers today. The problem of positivism that is relevant here is that when the current scientific understanding of a certain process is very small, positivist thinking tends to impose the theory onto the reality, paradoxically creating another kind of metaphysics, which, while expressed with logical deduction, has a very loose relationship with the reality. While this process probably inflicted most damage in the social sciences, natural sciences were also seriously affected. A typical positivistic distortion is the transformation of Lamarck's adaptation theories into a rigid interpretation of the relationship between form and function, where it is postulated that since nature is perfect, and organisms' form is optimal to their functions, from the study of their form we can unravel their functions. Of course this is pure metaphysical thinking, much closer to religious belief than to science, although it uses logic deduction or mathematical models.

In this complex cultural milieu, in 1867 a German anatomist, von Meyer, reported in detail the spatial organization of trabeculae in the proximal femur, and he could see, in whole bone, sections along the frontal plane; a Swiss engineer, Culmann, noticed that this organization closely resembled that of the so-called stress trajectories he had calculated for a similarly curved structure, a crane, using a graphical statics method that he had invented. They agreed to publish the two drawing side by side in von Meyer's paper, one of the trabecular organization and one of the crane stress trajectories (Figure 6.4).

In 1892 Julius Wolff, another German anatomist, published *Das Gesetz der Transformation der Knochen* (*The Law of Bone Transformation*), where he suggested

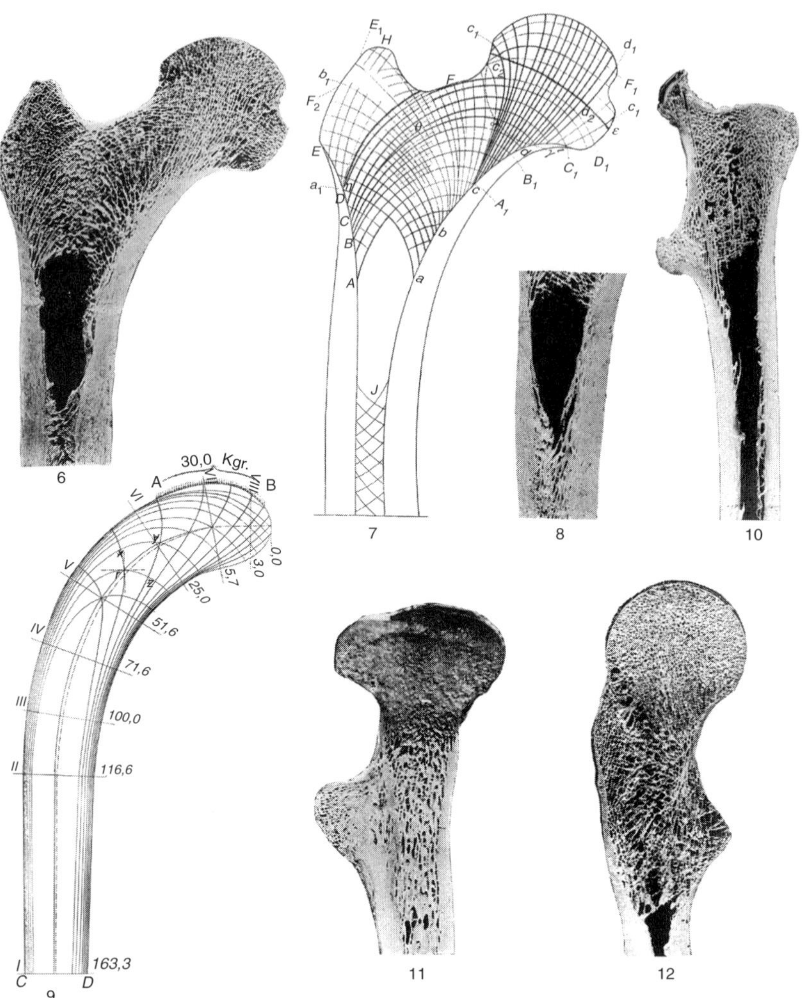

Figure 6.4   Wolff's reproductions of the original drawing, where Culmann's stress trajectories for the Fairbairn's crane are compared to the trabecular orientation in the proximal femur observed by von Meyer. Reproduced from: Julius Wolff, *Das Gesetz der Transformation der Knochen*, Hirschwald, Berlin 1892 – Reprint Pro BUSINESS, Berlin 2010. Reprinted with permission of the Julius Wolff Institute.

that the reason for the similarity observed by von Meyer and Culmann is due to the fact that bone adapts its shape and structure in response to changes in loading. For a number of reasons, this concept became a postulate in clinical practice, and today orthopedic surgeons still cite "Wolff's law" as the basis for understanding bone adaptation to loading. Wolff's work, as well as that of many of his followers, exudes positivism. Implicitly, there is the idea that the form and the architecture of bones is the optimal one with respect to the biomechanical function, and when this function changes, bone remodeling adapts bone, so that its form becomes optimized for the new biomechanical function.

From the 1960s, experimental biology became sufficiently advanced to enable some controlled experiments on the effect of altered biomechanical conditions on bones. Some results suggested that bone-tissue strain, as measured at the body–organ level, was a driving factor. Apparently, each bone had a desired level of strain, and if the external loading changed, the bone adapted so that the strain this new load induced in the tissue returned to the desired level. Some experiments showed that for small changes no appreciable processes of adaptation occurred; this was interpreted as a so-called "lazy zone" (also called a "dead" or "comfort" zone (Frost, 1997)), a range of strain values around the desired level that only activated the adaptation when exceeded.

In 1976 Stephen Cowin published his theory of adaptive elasticity, where he imagined a material capable of self-adapting its stiffness so as to maintain the level of stress and strain in each point at a preset value as the external loading changed (Cowin and Hegedus, 1976). By the 1980s, the finite-element method was sufficiently developed to compute stresses and strains induced in a bone-like structure by whatever loading, and computers were powerful enough to allow the implementation of some iterative schemes that simulated the various adaptation schemes that Cowin and other authors were proposing.

In the mid-1980s, what had been until then a research topic for a few bone physiologists became of extreme clinical importance. Thanks to the good results of the Charley hip prosthesis, in the 1970s, total hip replacement became a very popular treatment for the too many patients affected by osteoarthritis and rheumatoid arthritis. The components of the Charley prosthesis fitted the pelvic acetabulum and the proximal femur by the interposition of an acrylic polymer (called bone cement) between the metallic part and the bone tissue. However, some clinical studies showed that this polymer easily fractured over time, requiring additional surgery to re-stabilize the prosthetic components. Driven by these problems, some surgeons, such as Lord in France, developed so-called "cementless" prostheses, where the metallic part was designed to stabilize directly in contact with the host bone. The problem was that the femoral component, called the stem, was usually very long and thick and once the bone attached to its surface, the bending stiffness of the whole femur increased dramatically. The strain induced in the proximal femur bone by physiological loading was reduced drastically, and bone adaptation started to resorb bone around the prostheses. Naturally, this caused great concern, because seeing how quickly the bone disappeared in the control radiographs, and extrapolating this trend linearly, it was easy to conclude that in a few years all patients would develop spontaneous fractures around these cementless stems. In reality, as mentioned already, bone adaptation to underloading has a negatively asymptotic evolution, and indeed after a few years this process slowed down and stopped, and indeed very few spontaneous fractures were observed; but at the time this was not so obvious and the clinical community was very concerned. Suddenly, understanding (and possibly predicting) bone remodeling was of great importance.

A first wave of predictive models represented the bone remodeling process, as an optimization problem where a self-adapting structure changed its shape or density (stiffness) until the intensive quantity selected to represent the deformation (strain, strain energy density, effective stress, etc.) induced in the bone by the loading matched the homeostatic value in every point. This value could be defined a priori to be the same for every point of the tissue in the bone, or defined point-by-point by first assuming a "normal" loading, predicting the deformation in each point under that loading, assuming that as a homeostatic value, changing the loading, and then adapting the structure until the deformation returned to the homeostatic value at every point. These models were purely phenomenological, in the sense that they did not include any theory of how bone remodeling actually happened at the cellular level, but merely represented its effects at the organ level. As usual in these cases, researchers spent a lot of time arguing on which deformation indicator had to be used, and on technical aspects, such as the lack of stability in certain early numerical implementations. But it was only after a while that some research groups started to compare the predictions of these phenomenological models with experimental observations of bone adaptation in humans or in animal models.

The results were frustrating, to say the least. At best some authors showed a mild *correlation* between predictions and measurements, but no one got even close to actually predicting reality. This did not prove these models false; it simply showed that we lacked the ability to carry out such mechanobiology experiments. At that time no one was capable of generating a finite-element model of a bone capable of predicting stresses, strains, and displacements with errors of 10% or less from imaging data (CT, serial sectioning, etc.). Also, the methods used to "measure" bone adaptation were very crude. With uncertainties in the prediction of the basic quantities (strains) of 20–30% and in the measurement of the reference quantities (density changes) of 5–10%, it is obvious that trying to validate methods aimed at predicting changes in bone density that typically ranged between 3% and 30% was impossible.

One lesson to be learnt here (and a similar case was already discussed in Chapter 3, on prediction of muscle forces during movement) is that no serious modeling research can be done on a topic on which no accurate and detailed validation data are available. When it is impossible to establish what is right and what is wrong, modeling becomes a fairly futile exercise.

Over the last 20 years the field moved from purely phenomenological models to more mechanistic models based on the growing understanding of bone biology summarized in the previous sections.

In parallel, methods to predict stresses and strains in bones with subject-specific models improved dramatically, as well as the ability to create truly controlled mechanobiology experiments, and to measure accurately how these controlled conditions change the morphology and the degree of mineralization of the bone tissue. But in doing this awareness rose in the community: bone adaptation occurs at the cell–molecule scale, bone loading occurs at the body–organ scale. Bone remodeling is thus best modeled in between, at the tissue–cell scale. Thus, today most interesting developments on bone remodeling models, and on their validation, are taking place at this scale.

6.3.3.4 Mechanical signals at the tissue–cell scale

In the previous chapters of this book, I described the bone biomechanics at a much larger scale. But if we talk about mechanobiology, we should first try to understand the mechanical signals as seen from the dimensional scale of living cells. The effect of forces acting on bones is manifest at the tissue–cell scale as a deformation of the extracellular matrix. But what we want to know is how this signal reaches the cells that drive bone remodeling. To date we have not identified a mechanism that directly links the deformation of the matrix with the regulation of the cellular activity; in every theory there is always the mediation of another mechanism: deformation of the cell's cytoskeleton, micro-fracturing of the extracellular matrix, or acceleration of the interstitial fluids. In addition, since the extracellular matrix is piezoelectric, and since the interstitial fluids contain ions, the deformation of the matrix and the acceleration of the fluid produce electromagnetic phenomena.

For a long while this was the most obvious possibility for describing mechanotransduction. Forces deform the extracellular matrix, and the cells adherent to the matrix (lining cells) or entrapped in it (osteocytes) deform as well. Of course, this would not work for cells that are not adherent to the matrix, such as osteoclasts or active osteoblasts. But there is enough signaling between osteocytes, lining cells, and the rest of the cells in the system to imagine mechanisms of regulation even without direct deformation of the detached cells.

There is now solid evidence that load-bearing bones are subjected to microfractures of the extracellular matrix under physiological conditions. If the matrix is fractured, we can imagine that:

(a) The osteocytes that embed it are killed or at least stimulated by the rupture of some canaliculi that are crossed by the crack, which is most likely to produce a rupture of the osteocyte process lying inside those canaliculi.
(b) The crack advancement releases fragments of the matrix into the interstitial fluid, and that the proteins of the matrix fragments interact with the surrounding cells.

All porosities of the extracellular matrix (Haversian and Volksmann's canals, canaliculi, osteocytarian lacunae) are filled with interstitial fluid. When the matrix deforms this fluid is squeezed and accelerated. Fluid acceleration may interact with osteocytes and lining cells directly, by shearing the cytoskeleton; biochemically, through some transport mechanisms; or by interaction with ultrastructural elements of the cells (ciliae, filaments, etc.), which are extroflexed into the interstitial space.

There is now solid evidence that electromagnetic phenomena occur when the extracellular matrix is deformed. However, the effect that these very mild phenomena might have on bone cells is much less evident.

6.3.3.5 Modeling bone adaptation at the tissue–cell level

From an operational point of view, bone remodeling algorithms at the tissue level are not so different from those previously described at the organ level. The differences are technical, in relation to the different level of topological complexity involved, and conceptual, in relation to the different meaning that the various bone remodeling equations assume at this smaller scale.

From a technical point of view, the problem has already been discussed in Chapter 5. Bone tissue, especially cancellous bone tissue, has a morphological complexity that is much greater than that of any whole bone. This imposes special approaches for generating the numerical model, which must be capable of transforming data from a microCT imaging system directly into a biomechanical model of the tissue volume (so-called voxel-based methods). Another technicality worth stressing is that so far almost every bone remodeling algorithm at the tissue scale proposed in the literature works on a binarized dataset: in other words, the microCT data volume is subdivided into two types of tissue: bone and non-bone. Bone tissue is assumed to have a constant module of elasticity, whereas non-bone tissue is assumed to have no mechanical properties. In this idealization, bone remodeling manifests as a morphological operator that transforms voxels from bone to non-bone, and vice versa.

The most important difference with respect to organ-level models is that when bone remodeling is represented at the tissue scale, it is possible to assign mechanistic physiological and biological meaning to the various elements of the model, which links such models to the underlying knowledge on the bone remodeling biology and biochemistry. In particular, the apposition and resorption rates can be expressed as functions of the replication rate, differentiation rate, apoptosis rate, and level of activity of each primary cellular population. Individual cells are not modeled explicitly, but within a compartment it is possible to account for the relative density of each cellular type.

Another big difference in working at the tissue level is that it is possible to devise animal experiments where the level of control of the mechanobiology processes is much greater. As an example, I refer to a recent murine model developed at ETH Zurich by Ralph Müller and co-workers, where a mouse model, treated according to conventional biological protocols (knock-out, ovariectomized, calcium or hormones supplemented, etc.) is subjected to daily controlled loading experiments on a caudal vertebra. Two percutaneous pins are fixed to the two adjacent vertebral bodies, and external forces are applied through a controller (Figure 6.5).

Probably the biggest innovation of this protocol is the possibility to provide a fully spatially referenced time-lapsed 3D image of the region of interest at microCT resolution (Schulte *et al.*, 2010). Using high-resolution in-vivo imaging and registration algorithms, it is possible to obtain a detailed map of the changes in the bone-tissue morphology between two time points in the same animal at the very same location (Figure 6.6).

These experimental set-ups can finally provide a fully controlled environment, where a given region of the skeleton of a single animal is subjected to known biomechanical and metabolic conditions, and its remodeling is monitored over time. The availability of a baseline against which we can challenge our predictive models will surely move bone remodeling prediction out of the impasse that it has been stuck in for the last 40 years.

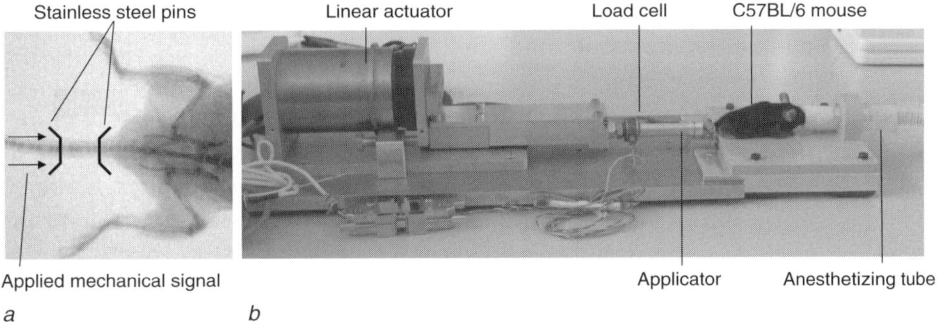

Figure 6.5 (a) Fluoroscopic image of a mouse, graphically edited to show the location and form of the stainless steel pins once they have been surgically inserted. The mechanical signal is applied to the distal pin whilst the proximal pin is clamped. (b) One loading axis of the caudal vertebra axial compression device (CVAD). Reproduced with permission from (Webster, 2010).

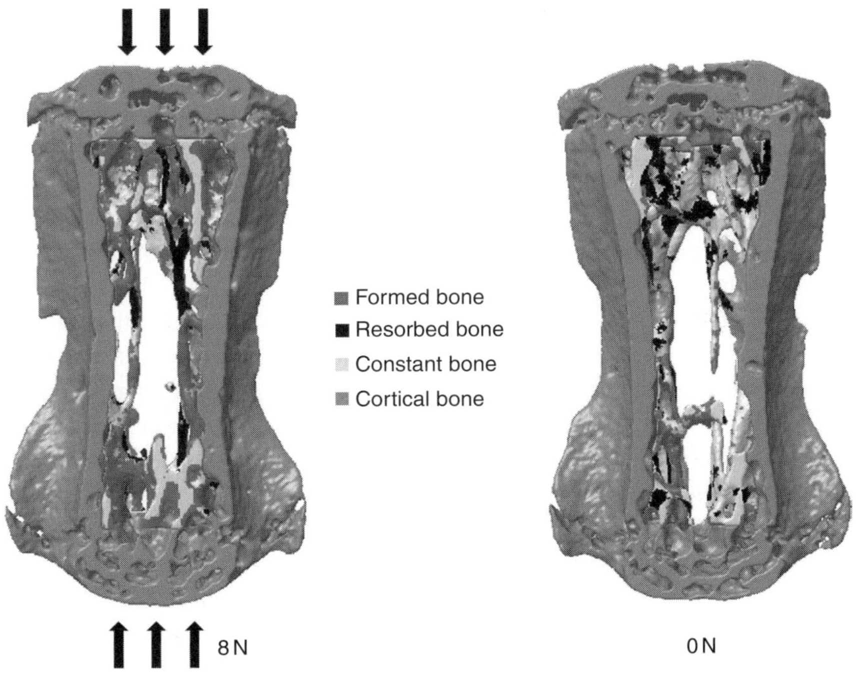

Figure 6.6 Three-dimensional visualization of formation and resorption sites in the trabecular compartment of a loaded (left) and a control (right) animal after four weeks. Areas of formation and resorption are shown. The loaded vertebra shows more formation sites than the control one, but resorption is still ongoing despite the mechanical loading regime. Reproduced with permission from (Schulte *et al.*, 2010).

6.3.3.6 Non-mechanical determinants of bone-mass regulation

So far, I have focused on the biomechanical modulation of bone remodeling. But this book would be a very bad example of integrative research if we were to stop there. As usual, bone metabolism is modulated concurrently by dozens of determinants, some innate and some acquired, including nutritional, hormonal, and lifestyle-related factors.

The first and most important determinant of bone-mass regulation is age. Even in the most ideal health conditions, bone mass decreases with age and the quality of bone changes, together with its functional attributes (van der Linden *et al.*, 2004). The skeleton undergoes dramatic changes over growth and then during aging, and neglecting this would be a gross error.

It has been suggested, from the results of a number of family and twin studies, that genetic factors may explain about 30–80% of the variability of bone mineral density (Rubin *et al.*, 1999). The huge range confirms that genetics is not in reality a direct determinant of the bone mass in adults. Of course, this does not mean that genetic determinants are not important, but only that their relation to the regulation of bone mass is complex, and modulated by a large number of acquired factors. The discussion over nature vs. nurture is old, and mostly pointless. Maybe we should imagine life as a single infinitely complex system of interactions over space and time, where the distinctions between the individuals and the environment, or between individuals and their offspring, are as blurred as those between tissue types or organ systems. To understand it a little with our poor, limited brains, we can decompose this unity for convenience (reductionism) but we should never forget the systemic nature of life. As we understand how complex bone-mass regulation is in terms of the number of biochemical species involved, we understand how many genetic determinants might actually play a role in bone-mass regulation.

Similar reasoning applies to nutrition determinants. While diets severely deficient in calcium or in the vitamins necessary to fix it (vitamin D) will certainly affect bone mass, eating a lot of dairy products and foods rich in vitamin D will not automatically ensure that we will preserve our bone mass during aging (Carins, 2005). More intriguing is the finding that energy metabolism affects bone remodeling. The findings suggesting that the leptin contained in fat might modulate bone mass through the nervous system suggest an endocrine perspective (Confavreux *et al.*, 2009; Chenu and Marenzana, 2005; Elefteriou, 2008; Karsenty, 2001). Moreover, bone secretes osteocalcin, a hormone pharmacologically active on glucose and fat metabolism. Thus, an endocrine control loop can be imagined, where cross regulation between bone and energy metabolism would place nutritional bone-mass regulation in a whole new perspective (Confavreux *et al.*, 2009).

Many hormones affect bone remodeling. I have already mentioned the parathyroid hormone, leptin, and osteoprotegerin. Estrogens are steroids that act as primary female sex hormones, in particular regulating the menstrual cycle. Among the other mechanisms, estrogens stimulate the expression of FasL, the ligand of a receptor that induces cell apoptosis when activated (Imai *et al.*, 2009). It is not known whether the regulation takes place by direct apoptosis of mature osteoclasts, or by apoptosis of pre-osteoclasts, which prevents osteoclasts from forming (Krum *et al.*, 2008). The role of the thyroid and its hormones is very complex, and not fully understood. Calcitonin, produced in human beings primarily in the thyroid, inhibits osteoclastic activity; a similar effect is produced by the thyroid-stimulating hormone.

Among lifestyle-related determinants, the levels of physical activity and of neuromuscular fitness were obvious candidates. However, meta-analyses of the literature do not provide strong conclusions. While there is a general health benefit in maintaining good levels of physical activity during adulthood, the direct effect of exercise on increasing bone mass is evident only during childhood, while in adults results are contradictory (Nevill *et al.*, 2003; Schiessl *et al.*, 1998; Skerry, 1997; Turner and Robling, 2005). Also, the idea that bigger muscles require bigger bones, while valid in growing subjects, lacks any conclusive confirmation in adults. Certainly, physical fitness, normal body weight, and good neuromotor conditions expose the skeleton to lower loading than is observed in pathological conditions, which in itself reduces the risk of bone fracture, even if the bone mass decreases with age. Other determinants, such as smoking or alcohol consumption, do have negative effects on bone mass that are documented by epidemiological studies, but the precise mechanisms that might link them with bone-mass regulation appear to be very complex, and are not fully understood.

7   Applications of multiscale modeling

Two real-world clinical applications of the multiscale modeling of the skeleton are described: one in pediatric skeletal oncology, and one in the prediction of the risk of fracture in osteoporotic patients.

7.1   Introduction

The scope of this chapter is to show, using a few clinical applications as supporting examples, how our ability to model musculoskeletal pathophysiology at different dimensional and temporal scales – as described in detail in the previous chapters – can be used to form multiscale models.

Whereas the previous chapters focused on explaining the physiological mechanisms that emerge at each scale, and the modeling methods involved, here we shall look at the practicality of composing multiscale models, indentifying the necessary set of parameters, and generating predictions that are of clinical relevance. Thus, our perspective will be much more methodological. This could condemn this chapter to a rapid obsolescence: the massive research effort on the development of the VPH framework of methods and technologies will change the way we do clinical multiscale modeling in the near future. Still, while the solutions (or their lack) described here will hopefully become obsolete, the problems to be solved will remain valid in years to come. Thus, I shall focus more on the problems than on the solutions, referring to specialized literature for the best practical approach to solve each problem to date.

The two clinical applications discussed are post-operative monitoring of pediatric skeletal oncology patients, and fracture risk assessment in osteoporotic patients. The first has been developed as an internal project at the Istituto Ortopedico Rizzoli, as a collaboration between our research group and the Skeletal Oncology clinical department. The second is the goal of the VPHOP integrated project, a large research endeavor funded by the European Commission as part of the first Virtual Physiological Human call for proposals of the seventh Framework program for Research and Technological Development of the European Union.

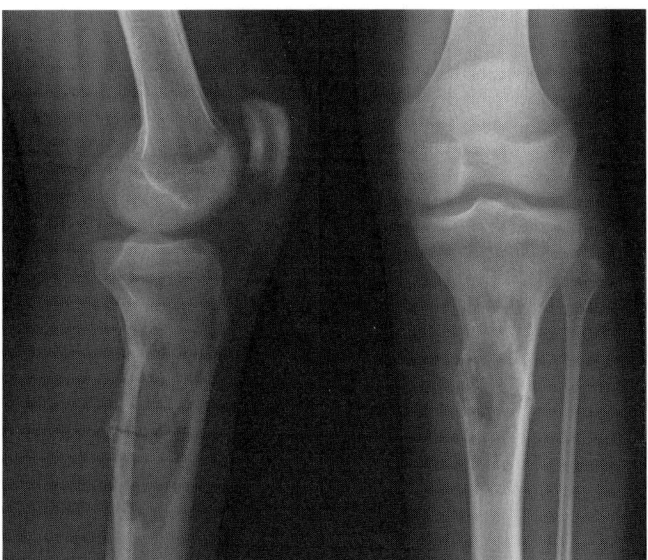

Figure 7.1   Anterior and lateral radiography of a large primary osteosarcoma affecting the proximal part of the left tibia of a 16-year-old male patient.

7.2   Post-operative monitoring of pediatric skeletal oncology patients

7.2.1   The clinical context

Malignant bone tumors (Campanacci, 1999) are rare diseases, with a strong prevalence in children and adolescents. This family of diseases includes chondrosarcoma, osteosarcoma, Ewing's sarcoma (Jedlicka, 2010), and various other sarcoma types (Figure 7.1).

In most cases, the tumor mass has to be surgically removed, and this requires either the amputation of the affected limb or, more conservatively, the resection of an ample portion of the affected bone (Marulanda *et al.*, 2008). In this second procedure, which became more frequent as clinical protocols improved, the surgeon must reconstruct the continuity on the skeleton after the affected region has been removed. In adults, this is done using prosthetic devices, metallic endoprostheses that are attached to the remaining stumps of the bone, and ensure the correct biomechanical function. However, in growing subjects, the rest of the skeleton will grow over time whereas the prosthesis will not, and this might create some severe problems. Since in long bones, a good part of the growth is in length, one possible solution is to design the prosthesis as a complicated telescopic device, whose length can be adjusted from a small percutaneous access or, more recently, with a remotely controlled magnetic actuator (Casas-Ganem and Healey, 2005; Nystrom and Morcuende, 2010). This approach has a number of limitations and various surgeons prefer the use of complex biological reconstructions instead of these devices, counting on the high osteogenic potential of children's skeletons. There are various

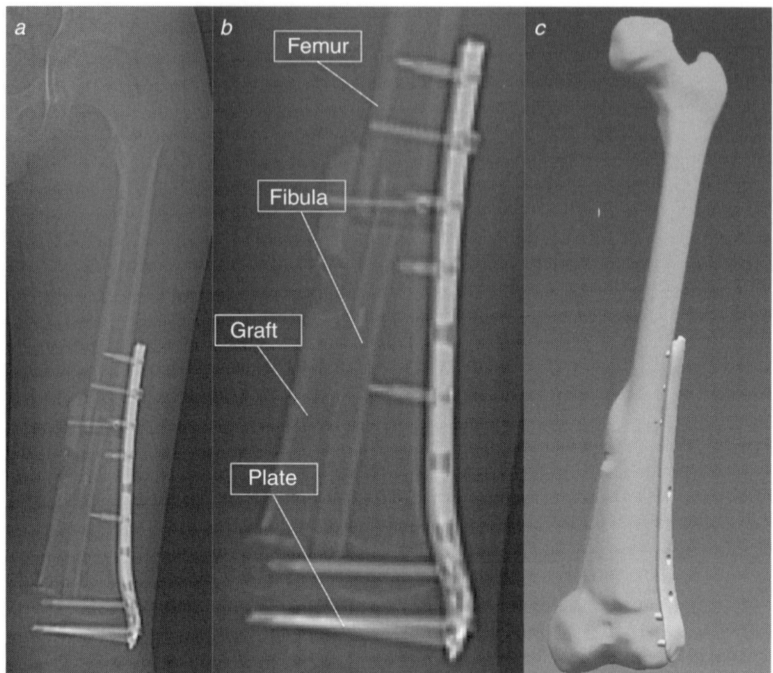

Figure 7.2   (a) Biological intercalar reconstruction of the distal femur in a child affected by an osteosarcoma. (b) Enlargement of the constructed region: the patient femur is resected, leaving the entire proximal femur and a small portion of the distalmost part, which includes the articular surface. The missing segment is reconstructed with a homologous graft from the bone bank, fixed with a metallic plate, and inserted as an auto-transplanted fibular autograft. (c) Computer model of the reconstructed bone and the plate as derived from CT data.

techniques, which in general involve the reconstruction of the portion of skeleton resected using a combination of bone-bank tissue (cadaveric bone); auto-transplanted vascularized tissue, which is removed from regions where this produces minimal functional damage, such as the fibula; and screws and plates used to attach these grafts together and stabilize them with the two skeletal stumps (Figure 7.2) (Ghert *et al.*, 2007; Innocenti *et al.*, 2009).

A big clinical problem with this procedure is that because of the complexity of the biological reconstruction it is very difficult to estimate how strong the reconstructed bone will be. After surgery, the reconstructed bone will be the site of intense healing, growth, modeling, and bone remodeling processes. The morphology and density of the various bone tissues (homologous, autologous) will change quite quickly, and again the effect of these changes on the ability of the reconstructed bone to stand intense loads will be difficult to guess.

However, knowing the loading capacity of the reconstructed bone is very important. These children face a long period of bed rest before and after the operation, in relation to the various therapies required. This lack of mobility produces severe negative effects on the regular growth of the child's musculoskeletal system; as soon as clinically possible, the child must enter a fairly aggressive rehabilitation program, aimed at restoring the musculoskeletal masses, to ensure the best possible functional outcome after such massive surgery. But tuning the intensity of the rehabilitation program is very difficult. If we are too aggressive too soon, we might exceed the strength of the reconstructed bone, and induce a fracture (which would require complex additional surgery and in some cases might even force an amputation). If we are too light, the child's growth and function might not be fully recovered.

The possibility of predicting the strength of the reconstructed bone, starting from medical imaging and biomedical instrumentation controls, as it changes over time as a result of the various biological processes would be a dramatic help in the effective management of these difficult clinical cases.

7.2.2 The multiscale problem

Even under physiological conditions, the anatomo-functional variability of children is very large, when compared with that of adults. In addition, the patients here underwent complex surgery that significantly modified not only the skeleton, but also the musculature in the operated region. Some muscles are damaged by the surgical access but, more importantly, owing to the ample skeletal resection some muscles have to be detached; some are re-attached at different locations, while others simply cannot be re-attached. Thus the musculoskeletal functional anatomies that emerge from these surgeries are unique, and each patient is totally different from any other. This imposes the use of highly personalized models, but also the need to account for the changes not only at the organ level but also at the whole-body level.

A personalized musculoskeletal body–organ model is necessary to determine the muscular and articular forces that are transmitted to the reconstructed bone during each physical activity (Viceconti *et al.*, 2006b). As we shall see, in reality this is already a multiscale model, as we couple an organ-level model of the contractile behavior of the single muscle with a whole-body model of the musculoskeletal dynamics.

The forces predicted by the body–organ model become the boundary conditions of the organ–tissue model, which represents the entire reconstructed bone, and in most cases also the contralateral intact bone, so as provide comparative results. It should be noted that in general the deformations of bones under load are so small that their effect on the body–organ level model would be minimal. Thus, the two models (body–organ and organ–tissue) are not coupled in both directions; we simple run the body–organ model, predict the forces, and impose these into the organ–tissue model.

The two models can be executed separately, using the most suitable software and hardware for each problem; in general their complexity is limited, and they can both be solved easily with a personal computer. The only detail one needs to account for is the application of the muscle and joint forces to the organ-level model. Given the perfect spatial registration between the bone geometry in the body model and in the organ model (as they are derived from the same 3D surface), when possible the best solution is to impose the points of application of these forces as defined in the body model as "hard points" in the organ model. This feature, today supported by most finite-element mesh generators, makes it possible to ensure that a finite-element node is placed exactly at a given location on the surface during the mesh generation. If too many hard points are placed, especially in regions where the surface has a high double curvature, the mesh generator can fail, or produce less than optimal meshes. However, in our case the number of points is quite small, and most are placed on relatively flat regions of the femoral surface, so this problem rarely appears. If the hard points are not available in the code you use, of if you encounter these mesh-conditioning problems, it is fairly simple to spread one force over the three corner nodes that delimit the element face on which the force application point falls. One only needs to make sure that the spread is chosen to satisfy both the force and moment equilibrium conditions.

Now let us consider each of these models in detail.

7.2.3   Body–organ model

A body–organ musculoskeletal model can be personalized starting from CT or MRI data (Cristofolini *et al.*, 2008). Ideally, both would be available. From CT data we can accurately reconstruct the bone geometries, whereas from MRI we can more easily reconstruct the action lines of the various muscle bundles, the muscle belly volume (which is necessary when a multi-fiber approach is used), and even estimate, although with varying accuracy, the physiological parameters required by the muscle contraction model (peripheral cross-section area, tendon and muscle rest lengths, pennation angles, etc.). When this is not possible, it is preferable to have the CT scan, which is mandatory for clinical monitoring of the post-operative phase and is also necessary for the organ–tissue level model. A good compromise is to have a full CT and MRI at the first control, and then to repeat only the CT at each follow-up control.

In the protocol currently in use at our institution, the patient's lower limbs are covered with skin-attached reflective markers that are also radio-opaque (Figure 7.3). To ensure that the landmarks do not move during the process, the whole limb is then wrapped with a transparent adhesive plastic film (Steri-Drape, 3M, USA). After landmark dressing the patient is examined with CT and MRI imaging. Since the imaging examinations are also required by the clinical protocol, the scanning plan must satisfy multiple requirements, and thus it is rarely optimal. However, generally speaking, since we need to derive geometry information from the images, we ask to use the thinnest X-ray beam possible (slice thickness), and to keep the pitch of the helicoidal scan as small as possible. Spiral CT reconstruction is usually performed at 1 mm slice distance.

If the MRI exam is also allowed, this is usually total body, so as to estimate the upper body mass more accurately as well. Various sequences can be used depending on the MRI model, but the goal is to maximize the contrast between muscles

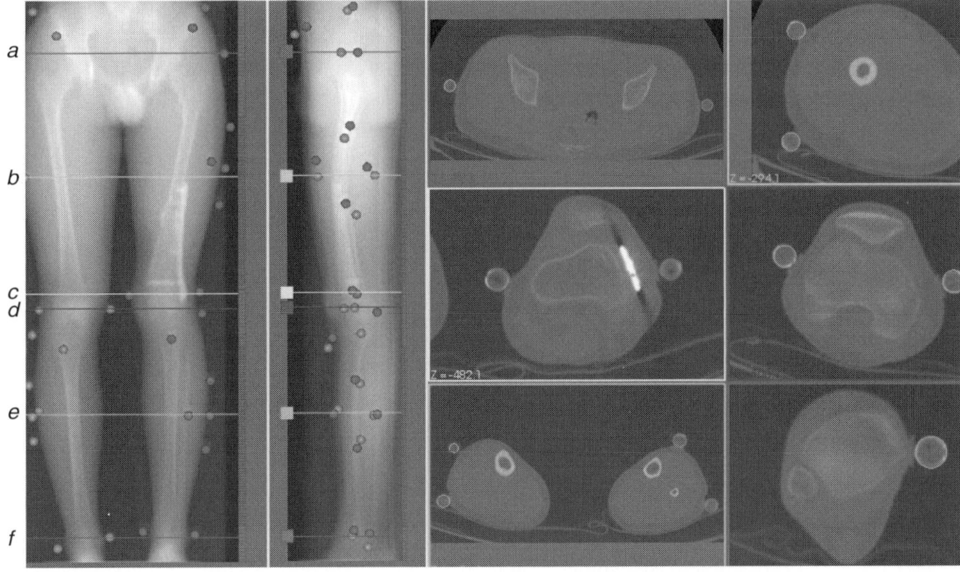

Figure 7.3   CT dataset of a pediatric patient who received a biological reconstruction of the right distal femur, following the resection of an osteosarcoma. Reflective radio-opaque markers attached to the skin can easily be identified in the CT images, providing a fiducial point set for registration with the motion capture data.

and the rest of the soft tissues. For example, in many machines spin-echo sequences provide good results.

The MRI data also provide the skin surface, which when subtracted from the bone surfaces, provides the volume of water-equivalent soft tissue. This volume is partitioned for each limb segment, providing the information that is required to build the inertia matrix of each member of the model.

The last step is a complete gait analysis session at our movement analysis lab. The reflective markers attached to the skin are kept in place, while some others are added, and all are tracked together with the ground reaction force and the electromyography signal of selected superficial muscles, while the patient executes a certain movement relevant for the rehabilitation protocol.

The muscle contraction model requires estimates of the maximum isometric force at its optimal length, the tendon slack length, and the fiber pennation angle for each muscle in the model.

The maximum isometric force is estimated indirectly, by computing the physiological cross-sectional area (PCSA) of each muscle, and then assuming a tetanic muscle stress that is homogeneous and constant for all muscles and equal to a value derived from the literature. Although in the literature this parameter is reported to assume a broad range of values, sensitivity studies suggest that its influence on the muscle forces predicted by optimization algorithms (see later) is moderate. The PCSA for most muscles can be estimated directly from images, assuming tabulated values for each muscle for the pennation value.

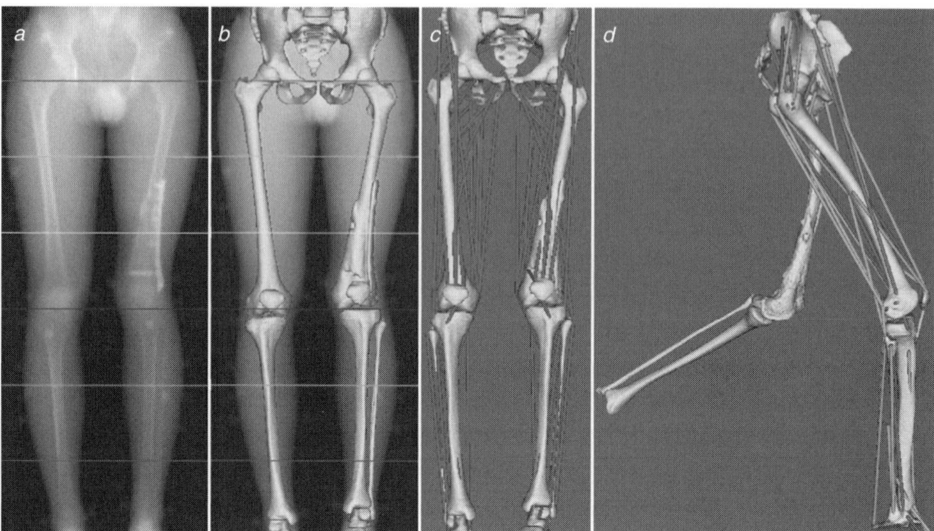

Figure 7.4   Construction of the whole-body model for the patient of Figure 7.3. (a) CT dataset; (b) 3D surface model of the lower-limb skeleton as segmented from the CT images; (c) Musculoskeletal model with the line of action of all muscles; (d) Using the skin-attached markers as a fiducial frame, the motion capture is registered to the musculoskeletal model and the inverse dynamic problem is solved.

All the length parameters are estimated, beginning with the muscle length in the supine posture of the imaging examinations; here, the sensitivity is not small, and there is room for improvement, possibly by using ultrasound imaging to capture the location of the insertions and of the aponeuroses that define the tendon length fraction of the musculotendinous complex.

The 3D geometry of the skeleton is well segmented from CT images (Figure 7.4b); software for data fusion makes it possible to combine and visualize the CT, the MRI, and the surfaces obtained from the bone segmentation simultaneously. This enables the construction of an accurate and personalized definition of the various linear actuators used to model muscles (Figure 7.4c). In this phase it is necessary to decide for each muscle bundle how many linear actuators must be used (which largely depends on the PCSA of the muscle, its pennation, and the size of its insertion areas); then for each linear actuator it is necessary to define the origin and the insertion onto the bone surfaces. The last step is the definition of the wrapping: as bones moves relatively to each other, muscles might wrap around them, curving their line of action. In musculoskeletal models, this can be solved by placing some geometric constraints (via points, wrapping surfaces) that cause the fibers to wrap during the bone motion in a more or less anatomical way.

Sensitivity studies confirmed that the number of parallel fibers used to represent a muscle is quite critical for the accuracy of the biomechanical prediction, at least up to a threshold value, above which adding further fibers no longer changes the predictions. Nowadays, the location, positioning, and wrapping of the fibers is mostly manual, and thus there is an operative limit to the number of fibers we can reasonably model; thus, in most cases we are forced to use many fewer fibers than those necessary. As the processing methods become more and more automated, we hope it will be possible to use tens if not hundreds of fibers for each muscle, drastically improving accuracy.

The last step in the definition of the musculoskeletal model is the creation of the kinematic joints, a set of equations that define how the motion of one bone is related to that of others concurring in the same articulation (femur, tibia, and fibula, in the knee articulation). In many orthopedic applications, where the attention is focused on the more or less pathological kinematics of a joint, it is evident that the models in use must capture the local joint kinematics with the highest possible fidelity. In the present application, however, the articular joints are important only to ensure an appropriate global synchronization between bone segments during motion. Thus, in such an application, it is perfectly appropriate to use relatively simple joint models, such as ideal ball and socket joints, or even a single-degree-of-freedom hinge for the knee. The only critical operation is the location of the center of the idealized joint, which must be placed in an appropriate anatomical location from a functional point of view. However, the availability of detailed medical images enables very accurate identification of the joint centers for hip, knee, and ankle.

Once the musculoskeletal model of the patient is complete, it has to be registered in space and time with the motion capture data (Figure 7.4d). In theory, the spatial location of the skin-attached markers, present in both the imaging and motion capture data, should provide a fiducial reference framework. In practice, these markers are not directly attached to the skeleton, but separated from it by a layer of soft tissue of varying thickness; thus, during motion their trajectory is only partially related to the motion of the underlying bones. This problem is solved by defining the initial configuration of the model by fiducial registration with the spatial position of the skin-attached markers in a so-called static capture, an initial capture of all markers obtained while the patient is immobile and standing. This initial registration is also used to re-orient the skeletal segments from the supine position in which they have been segmented in the CT images into the configuration they have in the erect position. In addition, the registration step defines the position and orientation of the whole musculoskeletal model in the gait analysis lab reference framework, within which the ground reaction force is also defined.

Next the spatial configuration of the model is obtained for each capture frame, by minimizing the quadratic distance of the skin-attached marker from a selected point on the bone surface, which is chosen by vicinity and anatomical connection using a global optimization scheme. The result of this process is the sequence of spatial configurations for the musculoskeletal model, one for each time frame of the motion, fully registered in space and time with the external forces (ground reaction) and with the EMG signals recorded from some muscles during the motion.

Now the inverse dynamic model is entirely identified, and can be solved, following the methods presented in Chapter 3. Since we usually deal with slow movements, static optimization is used to compute the joint moments as a sequence of instantaneous equilibrium conditions. For each frame, the muscle forces are then computed by finding the activation pattern that provides equilibrium to the articular reactions and minimizes the summation of the muscle stresses.

7.2.4   Organ–tissue model

From the CT scan data, we can also derive a full organ model, following the methods described in Chapter 4. The CT scan data are calibrated with a densitometry phantom, so as to have an accurate quantitative relationship between the Housefield units provided by the CT scanner, and the mineral density of the bone tissue we use to estimate the bone elastic modulus of each point of the bone.

Both operated and non-operated contralateral bones are modeled, so as to provide all biomechanical indicators predicted by the model in comparison with the same value for a normal bone.

Muscle and joint forces as computed by the body model are applied as boundary conditions to the bone model, which predicts displacements, stresses, and strains at each point of the bone, during the entire motion. Usually there is a single instant where the forces transmitted are maximal, and attention is paid to this time frame.

The risk of fracture is computed as a safety margin over the strain limit, assuming a failure criterion where the bone fails as a fragile material in tension, and as a ductile material in compression. In most cases, the most unfavorable condition is in tension and the fragile failure mode is predominant, in which case we use a simple maximum principal tensile deformation criterion, to decide whether or not the fracture occurs. If the fracture is more triaxial, in principle more complex failure criteria can be adopted, but there is always a problem of proper identification for failure criteria that involve multiple, hard-to-measure parameters.

In pediatric skeletal oncology applications, where the growth of the child and the intense remodeling of the reconstructed segment produce significant changes over time, this modeling exercise can be repeated periodically, to ensure that the rehabilitation program develops within reasonable safety margins.

7.2.5   Verification and validation

The body–organ model is formulated as an ordinary differential equation (ODE) problem coupled with a minimization problem (to compute muscle forces from joint moments). In general, verification is not perceived as a major problem in these types of simulation, since the time variations are usually smooth, and this ensures reasonably good numerical conditioning. However, as usual, some caution must be recommended. For example, we recently found a case where changing the tolerance of the de-nosing filter that we applied to the ground reaction force signal changed the hip reaction force. Of course, this should not happen; investigation revealed a peculiar interaction between convergence tolerances of the muscle force prediction algorithm with that of the ground force filtering. All numerical methods that involve a tolerance should always be explored, first checking the sensitivity of this tolerance on the results, and then estimating the uncertainty on the final prediction produced by a given value for such tolerance.

The second type of verification on the body–organ model concerns anatomo-functional constraints that are not explicitly imposed in the model. The presence of tetanic activation of single bundles during the simulation of low-intensity activities, or the prediction of joint reactions that exceed the normal physiological range that has been measured experimentally, usually suggests that there might some problems in the model predictive accuracy.

The true verification of body–organ models is still an open challenge. As the primary goal of this model is to predict the muscle forces, the most direct validation would be to measure such force in vivo. However, no experimental method currently allows this non-invasively. An indirect validation in the case under discussion is obtained by comparing the muscle activation pattern predicted by the body–organ model over the entire motor task with the electromyography recordings obtained from the same patient performing the same motor task.

For the organ–tissue model, there is a more solid verification and validation framework. In addition to the convergence test to ensure the adequate level of mesh refinements, we also use post-hoc indicators to ensure that the results are numerically accurate. Sensitivity analysis studies confirmed that the model did not amplify any of the experimental uncertainties affecting the input parameters. Last, but not least, extensive validation studies using cadaveric bones confirmed that the CT-based patient-specific model is capable of predicting surface strains and global displacements under known load with an average accuracy of 90% or better, and the load to fracture with an accuracy close to 90%.

7.2.6 Example results

7.2.6.1 Proximal femur reconstruction

So far, this complex protocol has been used to investigate only a very small number of patients, owing to the time, cost, and complexity involved. Here I report data from two cases. The first is that of a 5-year-old female who had the whole proximal femur resected owing to Ewing's sarcoma. This situation is considered the most difficult from a surgical point of view, because the surgeon must also reconstruct the articulation. The biological reconstruction, performed at the Istituto Ortopedico Rizzoli, is shown in Figure 7.5. A massive homologous graft is used to reconstruct the metaphyseal part of the proximal femur; this graft is internally augmented by the proximal segment of the fibula, resected from the same leg with the growth plate and one major vascular pedicle intact. The latter is reconnected to one of the femoral arteries by a vascular surgeon. A plate is used to stabilize the reconstruction on the stump of the resected femur, re-establishing skeletal continuity. The most delicate part is the induction of a greenstick fracture in the portion of the fibula protruding from the graft, in order to reproduce the femoral neck curvature.[1]

---

[1] http://en.wikipedia.org/wiki/Greenstick_fracture

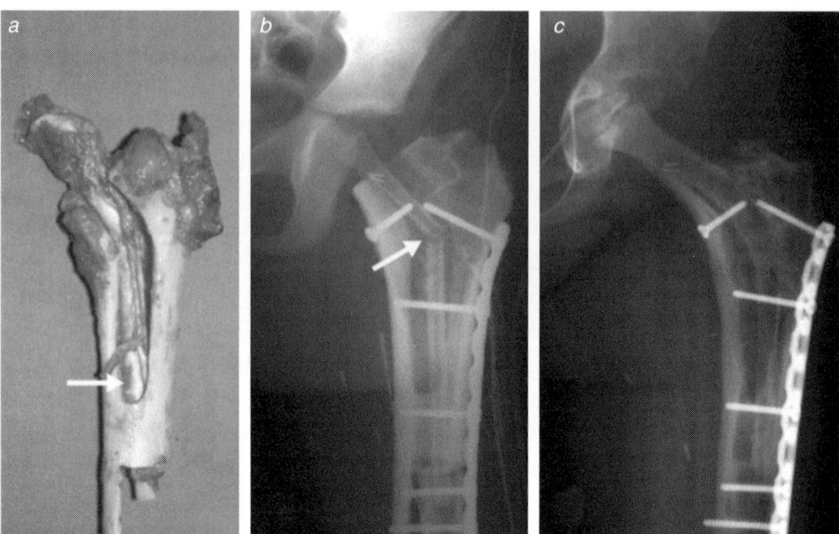

Figure 7.5   Implant prepared for reconstruction and radiographs of reconstructed hip. (a) Proximal femur: a massive allograft was reamed and shaped to hold the proximal fibula autotransplant, and the anterior tibialis bundle was anastomosed to the profunda femoris bundle in the recipient's thigh (arrow). (b) Post-operative radiograph; the fibula was fractured (arrow) to maintain periosteal continuity and to obtain a physiological cervicodiaphyseal angle. (c) Radiograph at 52 months after the operation. Reproduced with permission from (Manfrini *et al.*, 2003).

The patient was examined after the operation with the patient-specific modeling protocol described previously, in order to understand what the intensity of the rehabilitation program could be without risking a fracture of the reconstruction. Special care was used to extract the geometry of the metal plate from the CT scan data, and to model it and its fixation screws with the necessary detail.

The movement analysis showed a dramatic asymmetry during locomotion, partly due to the massive intervention on the proximal hip muscles, and partly due to the protective posture that the patient held during every motion. The study also highlighted an ancillary problem: since the fibula head is much smaller than the pelvic acetabulum, the reconstructed articulation did not show pure rotational kinematics. This aspect was further investigated in an additional study, using CT of the hip region in supine and flexed positions, from which we derived the roto-translational kinematics of the reconstructed articulation (Taddei *et al.*, 2005).

With all these data, we predicted the muscle and joint forces transmitted to the reconstructed femur and to its contralateral pair during slow level walking. Since the patient's primary rehabilitation work-out in the early stages would have been to walk unprotected in a pool with water up to her navel, we also simulated the reduced gravitational pull for different levels of the water.

The final results showed, as one would expect, a drastic difference in the strain field during walking between the reconstruction region and the same region in the contralateral femur. The area with the highest strain was that of the "femoral neck," i.e. of the portion of the fibula distal to the growth plate. The stresses in that region were high but not enough to predict a bone fracture (Taddei *et al.*, 2003).

Unfortunately, none of us remembered that in the same region we also had a much softer material, the cartilage of the growth plate. The thickness of the plate was comparable to the resolution of the CT images, so it was physically impossible to model it from the available data. As predicted, when the rehabilitation program started the child had no bone fracture, but after some time she developed a condition equivalent to the slipped capital femoral epiphysis (SCFE): the portion of the fibula proximal to the growth plate slid tangentially under the action of the shear forces. In an additional intervention, the surgeon performed a realignment of the fibular segments, and a fusion of the growth plate. After this episode, the child returned to her rehabilitation program, which was completed successfully.

7.2.6.2 Distal femur intercalary reconstruction

The second case I would like to describe is that of a 10-year-old boy, who had a distal femur intercalary reconstruction for an osteosarcoma. In these cases the surgical technique is slightly easier, as it is possible to salvage the most distal region of the knee articulation and of the relative growth plate. The post-operative simulation did not highlight any major problem, but during the second control we noticed an excessive stress shielding in the region between two screws. The surgeon found this result relevant, and having observed a successful stabilization, removed one of the two screws soon afterward through a small transcutaneous incision.

This patient was followed-up for four years after surgery, with the aim of monitoring the long-term modeling and remodeling that such massive reconstruction involved. The entire evolution of the risk of fracture in the two femurs over the follow-up period is represented in Figure 7.6. The length of each model was kept in scale, to show how the length of the bones changed over this time with the skeletal growth of the child.

As far as I know, this is one of the few examples reported in the international literature where patient-specific multiscale modeling has been used to produce clinically relevant recommendation for individual patients.

7.3 Predicting the risk of low-energy bone fractures in osteoporotic patients

7.3.1 The clinical context

Osteoporosis is a pandemic. These are a few facts and figures collected by the International Osteoporosis Foundation,[2] which also provides the bibliographic sources:

---

[2] www.iofbonehealth.org/facts-and-statistics.html

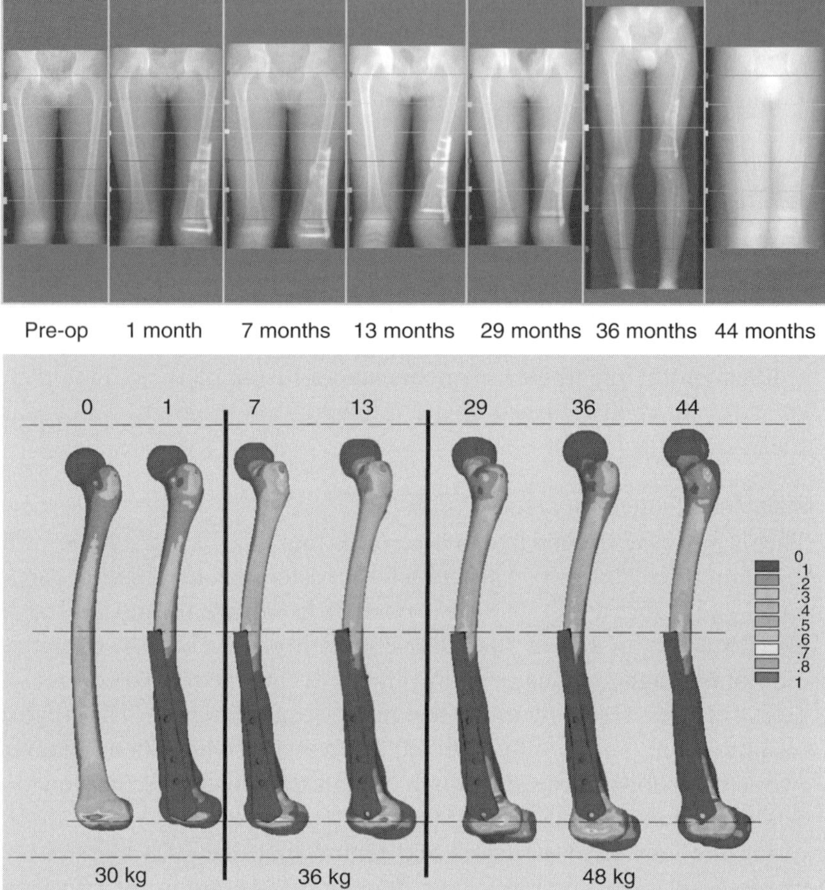

Figure 7.6 Above: series of control CT scans of a single patient, up to 44 months after the operation. Below: evolution of the risk of fracture over time in the same patient. The growth of the child in this period is evident by the changing length of the femur and the increase of body weight, indicated at the bottom.

- Osteoporosis affects an estimated 75 million people in Europe, USA, and Japan.
- One in three women over 50 will experience osteoporotic fractures, as will one in five men.
- About 30–50% of women and 15–30% of men will suffer a fracture related to osteoporosis in their lifetimes.
- By 2050, the worldwide incidence of hip fractures in men is projected to increase by 310%, and in women 240%.
- The combined lifetime risk for hip, forearm, and vertebral fractures coming to clinical attention is around 40%, equivalent to the risk of cardiovascular disease.
- A prior fracture is associated with an 86% increased risk of any fracture.
- In women over 45 years of age, osteoporosis accounts for more days spent in hospital than any other disease, including diabetes, myocardial infarction, and breast cancer.

- Evidence suggests that many women who sustain a fragility fracture are not appropriately diagnosed and treated for probable osteoporosis.
- It is estimated that only one-third of vertebral fractures come to clinical attention and under-diagnosis of vertebral fracture is a worldwide problem.
- Falls contribute to fractures – 90% of hip fractures result from falls. A third of people over the age of 65 fall annually, with approximately 10–15% of falls in the elderly resulting in fracture; almost 60% of those who fell the previous year will fall again.
- In 2000, the number of osteoporotic fractures in Europe was estimated at 3.79 million, of which 0.89 million were hip fractures. The total direct cost was estimated at €31.7 billion, which is expected to increase to €76.7 billion in 2050, based on the expected changes in the demography of Europe.

Nearly four million osteoporotic bone fractures cost the European health system more than 30 billion euros per year. This figure could double by 2050. After the first fracture, the chances of having another one increase by 86%. The risk of death related to hip fractures is as high for a woman as her risk of dying from breast cancer. We need to prevent osteoporotic fractures. The first step is an accurate prediction of the patient-specific risk of bone fracture that considers not only the skeletal determinants but also the neuromuscular condition.

The prediction of the risk of bone fracture is, in essence, a quite straightforward exercise. Human bones, like any other solid body, can only withstand a certain amount of external force before they fail. For a particular bone, this is determined by its strength and if we can know this actual strength and the forces likely to be acting on it, we can predict the probability that the bone may fracture now (actual risk of fracture). If we also know how bone strength and loading are going to change in the near future we can predict the cumulative risk of fracture over that period of time (absolute risk of fracture).

If we examine how fracture risk is currently assessed in clinical practice, we find some differences. Currently, the gold standard is to measure the mineral content in the bone. While this is one indicator of bone strength, it is clearly not sufficient. How can you decide if a bridge will stand, if the only information you have is the amount of steel used in its construction?

Because of this, the World Health Organization embarked on an ambitious project: the development of a clinical protocol for the assessment and prognosis of the risk of femoral neck fracture over ten years (Kanis *et al.*, 2007). This is a phenomenological model based on clinical observations of risk factors in large cohorts of patients. This model factors in age, body weight, lifestyle factors (e.g. smoking, activity), endocrinological factors, metabolic factors, etc. The multi-factorial model is very complex, much more so than direct assessment, as previously described. All of these external factors influence the bone properties and the applied loads that surely produce an effect on the final risk of fracture. To return to the bridge analogy, this is similar to a situation where to predict whether a particular bridge will crumble down under certain traffic, we observe many bridges similar to our own, exposed to different traffic scenarios, and see which ones fail, and which do not. Of course this works only if all the bridges are similar. If I plan to use a new alloy that was never used before, all this huge body of observations would become useless. The same applies to the WHO epidemiological model, called FRAX: if, for example, we try to predict the risk of fracture in patients who have already been treated with anti-resorptive drugs that slow down osteoporosis, FRAX cannot be used.

7.3.2   The multiscale problem

The problem of a mechanistic prediction of the risk of fracture in osteoporotic patients is that in reality this is a multiscale problem (Viceconti *et al.*, 2008b). Fracture occurs at the organ level, but the direction and intensity of loading is determined at the organism level. But even a body–organ / organ–tissue multiscale model, like the one used in the pediatric oncology application, would fall short here. Indeed, what we need is the ability to predict the chances that the patient will face a spontaneous fracture during the next five or ten years. To this end, we need to account not only for how the bone is today, but also for how it will change over time, owing to the progression of the disease, changes in other conditions, drug treatment, etc. But we now know that bone adaptation happens at the cellular level, and it manifests itself through morphological changes in the bone tissue. Thus, to be clinically relevant, a patient-specific model should account for phenomena that happen at the cell–molecule, tissue–cell, organ–tissue, and organism–organ scales.

There is a general problem in the application of personalized multiscale modeling methods to the prediction of the risk of osteoporotic fractures that we did not face with the pediatric oncology application. In that application, we were dealing with a patient affected by a life-threatening disease, who already in the standard clinical treatment was periodically examined with CT scanners; the ethical rationale is that the risk associated with the ionizing radiations is negligible in comparison to the benefits that the patient would receive if we can cure the cancer. Similarly, pediatric skeletal tumors are rare diseases with a dramatic social impact, which justify the use of complex, and expensive protocols.

In osteoporosis, we are dealing with a disease that is not immediately life-threatening (although the death rate associated with femoral neck fractures is impressive), that involves a significantly large portion of the population, and that affects mostly the elderly. This places much tighter ethical and economic constraints on the technologies and protocols we can use. This problem is being tackled in a European research project called VPHOP, which at the time of writing this book (2010) is running its second year of a four-year research program.[3] In this project, the consortium of research institutions is exploring various diagnostic technologies, with the fundamental idea of using progressively more invasive and more expensive technologies as the prediction of the risk of fracture suggests higher risk. Essentially, in screening where we have to monitor the healthy population for the insurgence of the disease, we aim to use technologies that involve very small costs and negligible

---

[3] www.vphop.eu

risk for the patient, at the cost of predictive models of moderate accuracy. For patients at high risk, more accurate – but also more invasive – examinations will be used, in a three-stage progression of increasingly complex protocols.

Since the project is still running, I shall list here below some of the options that are being taken into consideration in VPHOP, or that have been explored by other research groups. It is only with a much more extensive research activity, and with full clinical assessment, that we shall be able to decide which of these options has the best risk vs. benefit and cost vs. benefit ratios.

7.3.3 Body–organ model

In theory, we could consider the same protocol for patients at very high risk that we use for pediatric oncology, but this would not make sense for two reasons. Firstly, the cost–benefit ratio would be too high to be clinically interesting; secondly, the modeling scope here is different. In the pediatric application, the goal is to predict the forces acting on the reconstructed bone during the rehabilitation exercise, on which, of course, we have tight control. But in the osteoporotic patients, the loads acting on their weakening bones during their daily life are out of our control. At most we can hope to predict a probabilistic loading spectrum that somehow represents the patient's average daily activity.

A first option is to use a totally generic musculoskeletal model, scaled on the basis of the height and weight of the patient, to predict the average forces acting on the skeletal region of interest (typically the hip and the lumbar spine) for different motor tasks, and then to weight these various loading patterns with a probability function that somehow describes the patient's lifestyle and general neuromotor control condition. The lifestyle and the neuromotor condition can be estimated during ambulatory examinations and interviews, or with wearable sensors, some of which are designed to be minimally intrusive (e.g. hidden in a belt buckle) and can be worn continuously by the patient for days, capturing the lifestyle statistics but also recording spikes in acceleration or posture, that might suggest neuromotor problems (Daumer *et al.*, 2007).

A more advanced approach involves the use of 3D reconstruction from low-dose double radiographic projections to generate a patient-specific skeletal model, to which it is possible to scale an atlas of anatomical landmarks to define the insertions of muscles and ligaments (Chaibi *et al.*, 2011). This model is then used to predict the muscle forces during various conventional activities, which are weighted in the probabilistic load history as mentioned already.

7.3.4 Organ–tissue model

Once the body–organ model has been used to compute the probabilistic load history, this can be assumed as the probabilistic boundary condition for the organ–tissue model. Again, the organ–tissue model can be identified with increasing accuracy but also increasing invasiveness and complexity. Various researchers are working on the possibility of using single or double radiographic projections to initialize a parametric geometric model of the femur or of the lumbar spine. This parametric model can be obtained by principal-component analysis (PCA) of a large database of normal anatomies, all fitted with the same template geometry morphed elastically to each anatomy. The PCA provides a modal representation of the regional skeletal geometry; we can then use this modal representation in synthesis, generating the geometry that best fits the radiographic projections of the patients (Grassi *et al.*, 2010).

The alternative is to use CT images, with the same procedure as described for the oncology application. Here the intention is to keep the effective radiation dose as low as possible, as the imaging would be performed only for modeling purposes and not for other diagnostic or clinical monitoring reasons, as in the oncology application.

7.3.5   Tissue–cell model

In this chain of scales, the body–organ model defines the loading spectrum, and the cell–molecule model defines how the bone changes over time. In our group, we decided to pay greater attention to the organ–tissue model, which makes it possible to connect the loading spectrum with the strength of the bone, although this is disconnected from the cellular activity.

Other groups choose to sacrifice the possibility of defining the loading spectrum in exchange for being able to model the biological processes. In these protocols the patient is imaged with very-high-resolution CT scanners that provide a reasonably accurate reconstruction of the tissue morphology in a small region of the skeleton (Figure 7.7) (Melton *et al.*, 2010; van Lenthe *et al.*, 2008). These models are used with conventional loads, as the scope is not to predict the risk of fracture, but to estimate the bone strength. The possibility of having detailed tissue morphology then makes it possible to execute cell–molecule models that predict how this morphology changes as a function of cellular activity. The limit with this approach is that currently the only imaging methods capable of producing such high resolution in vivo involve high radiation doses, making these methods ethically acceptable only when used in the peripheral regions (wrist and ankle).

Another possibility being explored is to refine the organ–tissue level model in selected regions with higher resolution imaging methods, which could provide a sort of meso-scale spatial resolution (typically 150 micrometers) at a dose level that is acceptable in most skeletal regions.

7.3.6   Cell–molecule model

A variety of bone remodeling models are described in the literature. Some couple the cellular activity directly with changes at the organ level (typically apparent bone density), but most interesting models describe the interaction between bone multicellular units and the tissue, essentially predicting how the tissue morphology will change over time under the modulation of diseases, hormonal levels, the provision of bone drugs, etc. (Muller, 2005).

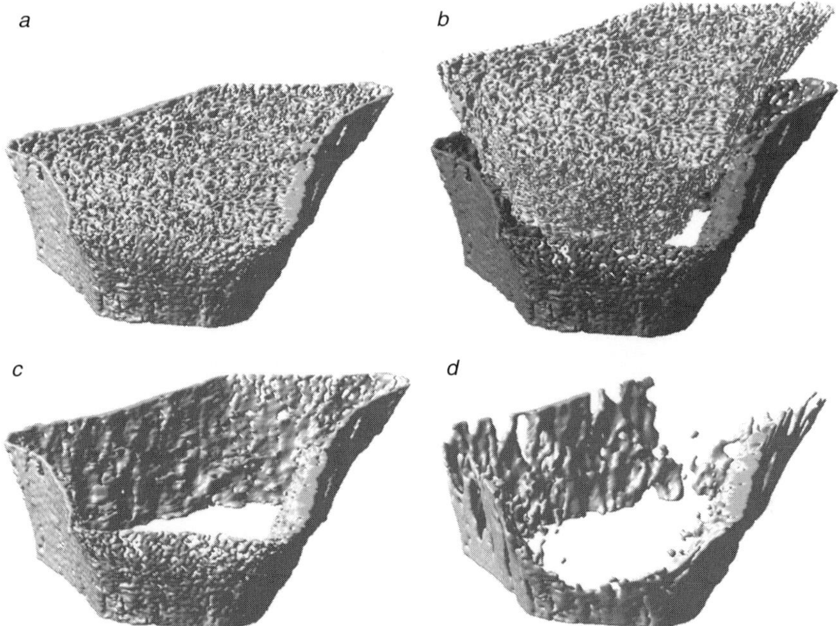

Figure 7.7 High-resolution 3D image of a portion of human radius generated by a HR-pQCT scanner in vivo. (a) Full compartment including both trabecular and cortical bone; (b) exploded view of the trabecular and cortical compartment; (c) typical cortical compartment as used in this study, which defined the cortical shell by an outer and inner hand-drawn contour; (d) the cortical compartment as defined by the automated procedure as recommended by the manufacturer. Reproduced with permission from (Mueller *et al.*, 2009).

7.3.7   Preliminary results

As mentioned above, we are describing here a work-in-progress within the **VPHOP** consortium. At present, the technology that would make it possible to run all models together in a complete integrative multiscale model is not yet available. Thus, the single models were run separately, and their outputs combined into a large-scale probabilistic Monte-Carlo simulation at the organ level, aimed at predicting the risk of fracture in a general population of Italians aged above 50.

The body-level sub-model defines the boundary conditions (i.e. the forces) acting on the organ-level model. An 82-muscle model of the lower extremities was used within a probabilistic representation of the neuromotor control similar to that described in Chapter 3. The model was further parametrized over the subject's body weight and average tetanic muscle stress, which when used for whole-muscle models, as in this case, can be considered as an index of the degree of sarcopenia of the subject. In addition, we parametrized it over multiple repetitions of level walking and stair climbing, to account for intra-subject kinematics variability.

The organ-level model was generated from a large collection of CT scans of the hip regions of Italian patients in the study age range, with no anatomical

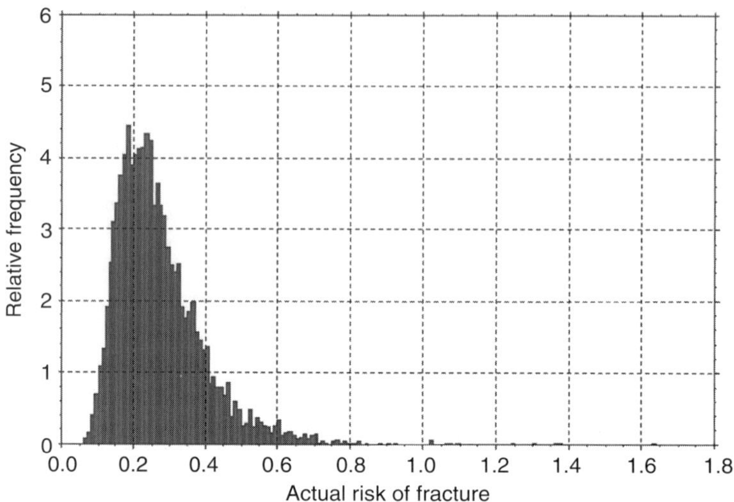

Figure 7.8   Frequency distribution of the actual risk of fracture over a simulated population of 50 000 subjects aged over 50.

deformation of the hip region. In addition to the femoral anatomy, a second probabilistic variable was provided by expressing the mineral density distribution as a function of the clinical hip T-score provided by conventional DXA measurements.

The tissue-scale model was used to generate at the organ level a constitutive equation that also accounted for the local anisotropy of the tissue due to the trabecular fabric orientation. The orthotropic components of the elasticity tensor, the Poisson's ratios, and the direction of the elasticity tensor were expressed probabilistically over a broad range of values.

Finally, the cell-level model was used to simulate how the parameters in the constitutive equation of the bone tissue would evolve over time, as a function of the subject age, and of the severity of the osteoporosis disease.

In addition to all these probabilistic inputs, all aleatory uncertainty due to the inaccuracy in measuring quantities was accounted for in the probabilistic model.

For a simulated population of 50 000 patients, the model predicted an incidence of fracture in the first year of 0.22% (figure 7.8). This result compares very favorably with the incidence of 0.27% observed in a recent epidemiological study (Piscitelli *et al.*, 2010), where the general Italian population over 45 was examined, using the national hospitalization database.

8   Multiscale modeling: the future

An analysis of the conditions required before the full integrative multiscale modeling vision can be embraced in full.

8.1   Introduction

In spite of its title, so far this book has mostly described how to model the human skeleton at a given scale. Only Chapter 7 described two clinical applications where models defined at multiple scales are combined to provide a reliable prediction of an otherwise complex process. But even in these examples, single-scale models were somehow patched together, without providing a clear general approach to multiscale modeling.

This is not a mistake, or an omission. The fact is, as explained in Chapter 1, that multiscale modeling is not a fully established reality yet, but rather a vision and a work-in-progress. The vision of representing human physiology and pathology as a concatenation of predictive and mechanistic models defined at different scales (the physiome) was proposed a while ago; but the concrete development of a framework of methods and technologies that make such a modeling approach feasible (the Virtual Physiological Human) has only just started, and, to date, none of the research projects that aim to develop portions of it has been completed.

The primary scope of this book is to survey how the biomechanics and mechanobiology of the human skeleton can be modeled at different scales using different methods and technologies. However, this book would be incomplete if I did not address the scenarios that are emerging in the development of the Virtual Physiological Human, both in terms of intermediate achievements, and of challenges still open. In spite of the risk of rapid obsolescence, this knowledge might lead the researchers interested in adopting multiscale-modeling approaches to viable and realistic methods, already within reach nowadays. Also, it might provide Ph.D. students some interesting targets in orienting their research programs and, ultimately, their careers.

In this chapter, I shall try to answer the following questions:

- How to identify multiscale models fully?
- How to transform multiscale modeling into an inquiry cycle?
- How to transform a model into a reusable quantum of (tentative) knowledge?
- How to create truly integrative models?

- How to falsify integrative models?

I can already predict that I shall answer them only partially; for the rest, I shall point the interested reader in what I consider the most promising directions, and highlight where the specific research challenges are.

8.2   Full identification of multiscale models

Let us for a moment assume that we already have the technology and the methods to generate mechanistic and predictive models of the human skeleton, ranging from the molecules to the whole body, and we intend to use them to make predictions relative to a given subject for whatever clinical purpose (prevention, diagnosis, prognosis, treatment planning, monitoring, rehabilitation). Such an integrative model would surely require a very large set of quantitative inputs measured on the subject at radically different space–time scales to be fully identified. The problem is, where do we get all those inputs? The problem of identification has been discussed in Chapters 3 to 6 in relation to each scale-specific modeling approach. Here, I just want to add some general considerations, which are valid for all components of an integrative model. Such a problem can be decomposed in four parts:

- Inputs that can be measured without any technical limitation: many of the inputs required to generate a detailed multiscale model of the skeleton can be measured without any particular technical or ethical issue. In such cases, however, one should never forget that if the predictive model we are developing is aimed for deployment in clinical practice there are also limitations that derive from the socio-economic and organizational impact that the introduction of such measurements in the routine clinical practice might involve. But such aspects are beyond the scope of this book, and will not be discussed here.
- Inputs that can be measured only while creating a hazard to the subject: when the measurement requires the subject to be exposed to a potential threat for his or her wellbeing (surgery, ionizing radiation, balance perturbation, etc.), the problem must be framed in an ethical context of risk–benefit analysis, and assessed by the appropriate ethical evaluation bodies.
- Inputs that are available only in qualitative or semi-quantitative form: except for very rare cases, the inputs of predictive mechanistic models must be quantitative. Unfortunately, very large bodies of observations, and some of the most established observational methods in medicine are totally qualitative, or semi-quantitative at most. In such cases we need to "re-invent" the observation protocol, transforming it into a fully quantitative exercise.
- Inputs that currently cannot be measured: there are quantities that at the current state of the art simply cannot be reliably measured.

In all cases, one can consider three alternatives to the direct measurement of a necessary input. The first is to estimate the input using reduced information sets, which is complemented by adding some a-priori knowledge, typically derived from a consistent population. A perfect example is the estimation of bone 3D geometry from one or two 2D projective images. This can be done (although with some serious inaccuracy) in various ways, but always using some reference atlas, some average 3D anatomy that is used as a template and transformed to best match the 2D images (Benameur *et al.*, 2005; Sadowsky *et al.*, 2007). The second alternative is to generate a *surrogate input*, an estimate of the quantity of interest, from a distribution of values measured in a group of model organisms for which we claim some sort of similarity with the subject being modeled, and for which direct measurement is possible. Such model organisms might be, for example, animals, human cadavers, or other human beings for whom the limitations preventing the measurement do not apply. The third alternative, mandatory for inputs that cannot be reliably measured at all, is to produce a surrogate input via indirect or inverse determination, i.e. by measuring another related quantity, and then to use another predictive model of the relation between the two quantities to estimate the value of the non-measurable quantity that would best match the measured quantity. A classic example is the identification of an intensive quantity, such as the modulus of elasticity of a tissue, by measuring an extensive quantity, in this case the maximum displacement under a known force, by solving an inverse elasticity problem, where we search for the value of the tissue elastic modulus that minimizes the difference between the predicted and measured displacement.

Replacing the measurement on the subject being modeled with a surrogate input always requires some caution. When the input of a predictive model is replaced by a surrogate input, the surrogate is a descriptive model, a property that under some idealization we pretend to represent that of our subject. As with any other model, before we can use it with some degree of confidence, we must try everything we can to falsify it.

At first we need to question the fundamental idealization: with respect to the property being measured, can we assume that the subject of interest is just a member of a population sampled by the measurements we obtain from the model organism? Is there any reason for which such property might be systematically different in my subject and in the group of model organisms I use to generate the surrogate input? For example: some properties can be measured only invasively; in most cases it would be considered unethical to perform a surgery just to take a measurement, but acceptable to take such measurements during surgery that is being performed for therapeutic reasons. This is an example where the input cannot be obtained for the subject being modeled because of limitations that do not apply to another group of individuals. However, if the model organisms are being operated on because of a disease that alters the property of interest, making them systematically different from that of all other subjects, including the subject that we are modeling, this would not work. When using human cadavers, we must be sure that the death did not alter the property of interest. When we use animal models, must check that the property of interest does not change considerably between species.

The second step is to understand how large the pool of surrogate measurements we use must be to generate our surrogate input, to ensure the necessary statistical significance and power.

The third step is to estimate how sensitive our predictive model is to the variability we observed in the surrogate input as measured over the model organisms. If such sensitivity is limited, then we can consider taking as our surrogate input a single synthetic indicator of the distribution of values we obtained from the model organisms (average, maximum value, 90th percentile, etc.). Otherwise, the uncertainty that the use of the surrogate input involves should be explicitly modeled as an aleatory uncertainty in the predictive model, which should be turned into a probabilistic model.

In many practical cases, these prescriptions may seem unrealistic; frequently the researcher uses a single measurement obtained by a third party on a model organism as a surrogate input, without any further questions. But we should try to follow them as much as possible; there is no point in developing a very sophisticated predictive model, and then feeding it with largely inaccurate inputs.

8.3   Transforming multiscale modeling into an inquiry cycle

It is my personal opinion that the biggest value in multiscale modeling in the long run is its potential ability to capture in full the inquiry cycle that represents scientific investigation. In Chapter 1, we noticed that abduction is a type of logical reasoning that makes it possible to infer the possible cause of an observation: according to pragmatic philosophers, such as Peirce or William James, the scientific inquiry process can be described in terms of logical reasoning as a sequence of abduction, deduction, and induction. Oversimplifying, abduction is used to generate a possible explanation from observations, deduction to transform this tentative explanation into a testable prediction (or a falsifiable hypothesis, according to Popper), and induction is then used to attempt a falsification of the explanation by comparing the prediction with other observations.

This concept works well to describe the process of building a predictive model, with the notable exception that in the practice of science we are rarely so lucky (or unlucky depending on the point of view) to have our tentative explanation completely and conclusively disproved. Mostly, our model predicts the observations we obtain from controlled experiments with some level of inaccuracy. By observing what is wrong, we frequently enter into a new abduction cycle, which pushes us into a new inquiry process: sometimes this is just a refinement or improvement of the current model; sometimes this leads us to whole new questions to be answered.

A perfect example of this continuous inquiry cycle is that where we expand the *explanatory horizon* of our model. Sometimes, the first attempt to build a predictive model involves a totally phenomenological approach, in which "black-box" modeling methods are used. Regardless of their predictive accuracy, which might be very high, these models will never tell us "why" something is happening; their explanatory content is null. So in most cases, sooner or later we end up with a mechanistic model, which provides an explanation to the causation we theorize. But no mechanistic model is entirely mechanistic: there is always an explanatory horizon, below which everything is still described in a phenomenological way. While this concept of the explanatory horizon is quite new and immature, it is quite clear that it is somehow related to the limits of validity of the model. But another way to see it is with respect to the characteristic space–time scale at which we formulated our modeling idealizations. In this context, a very common process in recent biomedical research is to recursively "drill-up" or "drill-down" to neighborhood scales. In 2002, Denis Noble wrote:

There has been considerable debate over the best strategy for biological simulation, whether it should be "bottom-up," "top-down" or some combination of the two. The consensus is that it should be "middle-out," meaning that we start modeling at the level(s) at which there are rich biological data and then reach up and down to other levels.

(Noble, 2002)

Is it possible to generalize this intuition into a modeling formalism? This is currently a point of debate. If we assume that there is a possible causal relationship between two portions of reality (Figure 8.1a), a possible way to represent the conventional modeling process (Figure 8.1b) would involve a descriptive model to transform observations into inputs for a deductive model, whose outputs (predictions) would then be compared with other observations through an inductive model. This static process could become a continuous inquiry cycle if we imagine that an abductive model uses the result of the comparison between predictions and observations to formulate an expanded explanatory horizon that revises the descriptive, deductive, and inductive model components (Figure 8.1c).

In particular, the expansion of the explanatory horizon could follow the middle-out approach that Noble and others recommend. We could start by modeling a phenomenon at the particular scale where most observations are available, or at the characteristic scale where the primary process that we intend to model is observable. So in the case of bone fracture this would be the organ scale, whereas in the case of the kinetics of an anti-resorptive drug that could be the cell scale. Then depending on how well our model is capable of explaining the observations at that scale, we might consider extending it to a different scale, above or below, actually creating a multiscale model.

In this sense, the creation of a multiscale model can be seen as an *abduction process*, through which we guess how to expand the explanatory horizon of our model.

8.4 Transforming a model into a reusable quantum of knowledge

8.4.1 Rationale

The problem of model reusability has only recently begun to be discussed. The traditional position that most have on this matter is that each model is conceived

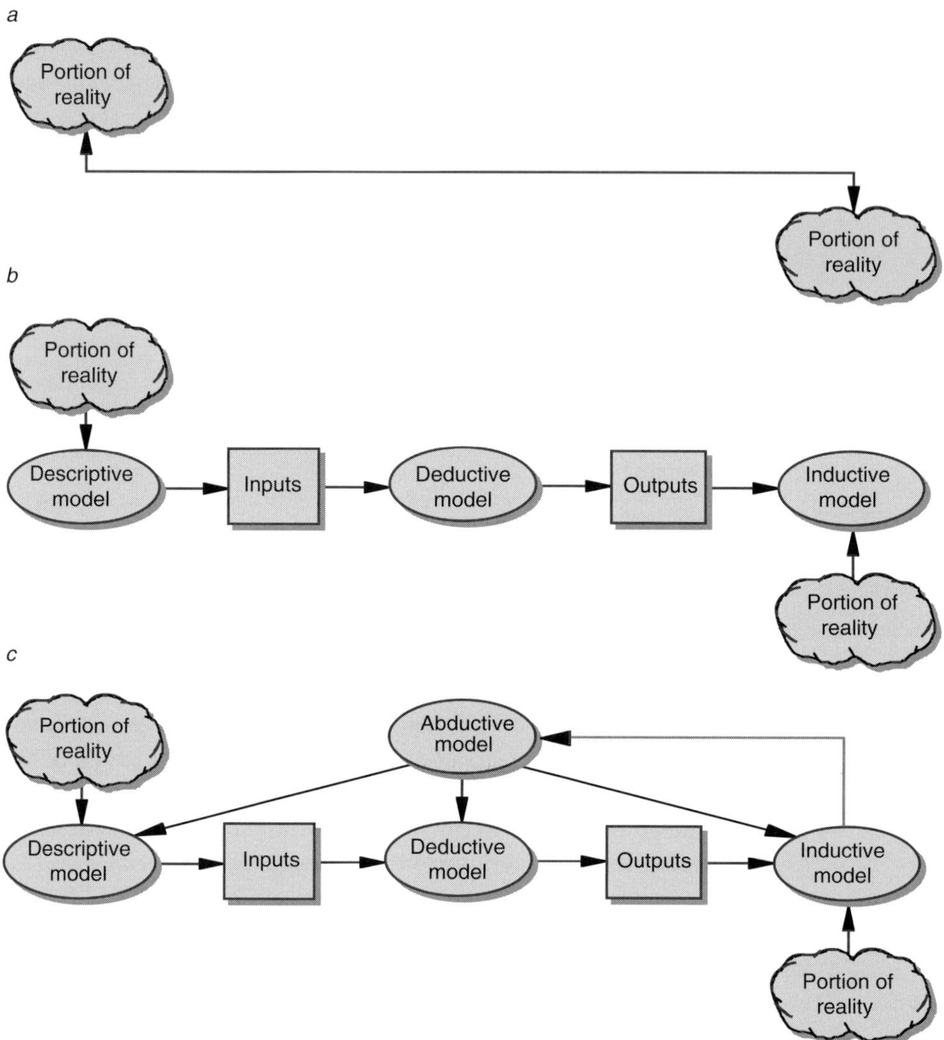

Figure 8.1 (a) Two portions of reality are in a potential causal relationship. (b) Conventional modeling generates a mechanistic explanation within a given explanatory horizon. (c) The inductive comparison between the prediction and the observation can inform an abductive model that updates the other modeling components by expanding the explanatory horizon of the model as necessary.

to answer a very specific scientific enquiry, which makes its reuse mostly pointless. This perspective changes once we accept the idea that a model may be a composition of multiple sub-models defined at radically different space–time scales. In this case, reusing a model could be an effective way to incorporate into our multiscale model the tentative knowledge developed by us or by others in regard to a specific scale.

Given that the idea might yield some usefulness, we are faced with the problem of how to make this possible. The problem is mostly technological in nature, and it has been formulated only recently (STEP Consortium, 2007) and probably not in full yet. But some aspects are already clearly outlined (Hunter *et al.*, 2010): the generation of a prediction using a model can be split up into the definition of the data we use as inputs for the model, the model itself, and the description of the simulation intended as the set of actions that produce the prediction. These three elements can be discussed in terms of minimal requirements, of standardized syntax, and of standardized semantics (Hunter and Viceconti, 2009).

8.4.2   Minimum requirements

There is a growing consensus that the biomedical research community should reach a consensus of what is the minimum set of information one should provide with one's experimental results in order to maximize their usefulness and to improve reproducibility of complex experiments. Probably the most important initiative in this sense is the "Minimum Information for Biological and Biomedical Investigations (MIBBI)" project (Taylor *et al.*, 2008).[1] In this context, the MIRIAM standard[2] (Le Novère *et al.*, 2005) attempts to provide guidelines on the "Minimal information required in the annotation of models." While the work of Nicolas Le Novère and of the other developers of the MIRIAM standard is mostly concerned with biochemical models, much of what MIRIAM regulates could be extended to any model. MIRIAM defines three aspects:

- *Reference correspondence*
  - The model must be encoded in a public, standardized, machine-readable format, and must comply with the standard in which it is encoded.
  - The model must be clearly related to a single reference description. If a model is composed of different parts, there should still be a description of the derived or combined model.
  - The encoded model structure must reflect the biological processes listed in the reference description.
  - The model must be instantiated in a simulation: all quantitative attributes have to be defined, including initial conditions.
  - When instantiated, the model must be able to reproduce all results given in the reference description within a given tolerance.
- *Attribution annotation*
  - The model has to be named.
  - A citation of the reference description must be provided (complete citation, unique identifier, unambiguous URL). The citation should enable identification of the authors of the model.
  - The name and contact details of the model creators must be provided.

[1] http://mibbi.org/
[2] www.biomodels.net/miriam/

- ○ The date and time of creation and last modification should be specified. A history is useful but not required.
- ○ The model should be linked to a precise statement about the terms of distribution. MIRIAM does not require "freedom of use" or "no cost."
- *External resource annotation*
  - ○ The annotation must enable a piece of knowledge to be unambiguously related to a model constituent.
  - ○ The referenced information should be described using a triplet, as: {data type, identifier, qualifier}.
  - ○ The data type should be written as a uniform resource identifier (URI).
  - ○ The identifier should be analyzed within the framework of the data type.
  - ○ Data type and identifier can be combined in a single URI, such as: urn:miriam:dataType:identifier. For example: urn:miriam:uniprot:P62158.
  - ○ Qualifiers (optional) should define the link between the model constituent and the piece of knowledge: "has a," "is a version of," "is homolog to," etc.
  - ○ The community has to agree on a set of standard valid URIs.

Similarly, the MIASE guidelines developed by the same team attempt to encode how whole simulations should be annotated.[3] The MIASE guidelines revolve around three principles:

(1) All models used in the experiment must be identified, accessible, and fully described.
  - ○ A description of the simulation experiment must be provided, together with the models necessary for the experiment, or with a precise and unambiguous way of accessing those models.
  - ○ The models required for the simulations must be provided with all governing equations, parameter values, and necessary conditions (initial state and boundary conditions).
  - ○ If a model is not encoded in a standard format, then the model code must be made available to the user. If a model is not encoded in an open format or code, its full description must be provided, sufficient to re-implement it.
  - ○ Any modification of a model (pre-processing) required before the execution of a step of the simulation experiment must be described.
(2) A precise description of the simulation steps and other procedures used by the experiment must be provided.
  - ○ All simulation steps must be clearly described, including the simulation algorithms to be used, the models on which to apply each simulation, the order of the simulation steps, and the data processing to be done between the simulation steps.
  - ○ All information needed for the correct implementation of the necessary simulation steps must be included, through precise descriptions, or references to unambiguous information sources.

---

[3] www.biomodels.net/miase/

○ If a simulation step is performed using a computer program for which source code is not available, all information needed to reproduce the simulation, and not only repeat it, must be provided, including the algorithms used by the original software and any information necessary to implement them, such as the discretization and integration methods.

○ If it is known that a simulation step will produce different results when performed in a different simulation environment or on a different computational platform, an explanation of how the model has to be run with the specified environment or platform in order to achieve the purpose of the experiment must be given.

(3) All information necessary to obtain the desired numerical results must be provided.

○ All post-processing steps applied on the raw numerical results of simulation steps in order to generate the final results have to be described in detail. That includes the identification of data to process, the order in which changes were applied, and also the nature of changes.

○ If the expected insights depend on the relation between different results, such as a plot of one against another, the results to be compared have to be specified.

Also, in the case of MIASE, the generality is such that these guidelines could be adopted for every kind of model.

8.4.3   Standardized syntax

For most biomedical data, there are one or more standards that prescribe how such data should be digitally encoded, stored, and transmitted. Probably the most important of these standards is the ACR-NEMA DICOM (Digital imaging and communications in medicine) standard, which is widely adopted in the area of medical imaging, but is now extending its scope, with the ambition of providing digital encoding standards for most non-textual clinical information. Another is the Protein Data Bank (PDB) format, which provides a standard representation for macromolecular structure data derived from X-ray diffraction and NMR studies. Where a domain-specific standard does not exist one can always resort to more general digital encoding standards, such as the Hierarchical Data Format (HDF5), which prescribes how to store and organize large amounts of numerical data. This is not to say that the standardization process for biomedical data is complete, but only to note how the process is in place and progressing. But the problem of interoperability remains, and the proliferation of data formats, standardized or not, only makes it worse. If we imagine that data flow from one model to the other, this means that data should be stored and transmitted in formats that each simulation program can read and write. Given the extreme diversity of the data that might be involved in a large multiscale simulation, the idea of creating a unique huge standard sounds quite unrealistic; it is probably more realistic to imagine the availability of translation libraries, which can be slowly extended to translate any relevant data format into any other alternative.

The story becomes a little more complicated when we talk about models. In the last few years, two opposite philosophies have been in constant conflict. The first, and most widely supported, assumes that it is always possible to separate the mathematical description of a model from the description of its numerical solution. Under this assumption, we can then imagine the development of a neutral language that can be used to encode models, which every simulation program supports. I can develop a model with simulation program A, store it in this neutral language format, and anybody else can download that model description, open it with simulation program B, and run it. Probably the most mature endeavor in this camp is the Systems Biology Markup Language (SBML),[4] followed by the CellML[5] and NeuroML[6] languages. Model repositories were established based on these mark-up languages, along with a long list of software tools to support them. All these mark-up languages are presented as very broad in scope, but in practice they are strongly characterized by the specific needs of the communities that developed them. So SBML is primarily a tool for biochemical modeling, CellML focuses on the combination of physics and chemistry models typical of cardiac modeling, and the NeuroML language focuses on neurobioelectric processes. But one feature that is common to all these endeavors is that they currently deal only with problems that can be represented using ordinary differential equations (ODE). This limitation is acknowledged; the CellML team is trying to address this with a new mark-up language called FieldML,[7] whereas the SBML community is exploring the extension of the SBML specification to include partial differential equation (PDE) models.

The second camp, much less organized and mature, questions the core assumption that for all kinds of model it is always possible to separate the mathematics from the numerics, especially when dealing with complex PDE problems. In addition, they stress the fact that the most advanced models also require special hardware to execute, be it a cluster, a high-parallelism machine, or a GPGPU system. To deal with these problems, they propose instead to store models as *remote procedure calls* (RPC), programs that accept inputs from a remote requestor, run, and return some outputs to the requestor. Typically, such programs would be published remotely as *web services*, procedures that can be invoked via hypertext transfer protocol (HTTP) and executed on a remote system. In such cases, it would not be important to know how the model has been encoded on the remote system, but only to standardize the communication to and from the remote system. The advantage of this approach would be that a researcher could share the model, its actual implementation, and the hardware used to solve it as a fully optimized service. It

---

[4] http://sbml.org/
[5] www.cellml.org/
[6] www.neuroml.org/
[7] www.fieldml.org/

should be noted that in this service-oriented scenario, the data format translation could also be seen as a remote service.

This split vision also affects how a standardized syntax for whole simulations could be provided. The group who developed MIRIAM/MIASE for this purpose propose another mark-up language, called Simulation Experiment Description Markup Language (SED-ML),[8] which describes a full simulation using five classes: the *model* class is used to reference the models used in the simulation experiment; the *simulation* class defines the simulation settings and the steps taken during the simulation; the *task* class combines a defined model and a defined simulation setting; the *datagenerator* class encodes the post-processing to be applied to the simulation results before output; the *output* class defines the output of the simulation.

The advocates of models such as RPC look at standardized syntaxes for services workflows, such as the Web Services Business Process Execution Language (WS-BPEL) OASIS standard,[9] or to full workflow management systems, such as the European Taverna,[10] of course, under the assumption that component models can be published as standard web services.

8.4.4   Standardized semantics

With respect to data, there are a number of available ontologies and metadata dictionaries. The DICOM standard has a full set of standardized annotation tags, while various groups are working on its full ontologization. The Gene Ontology standard[11] standardizes the representation of gene and gene-product attributes across species and databases. The BioPAX ontology[12] provides an abstract representation of biological pathway concepts and their relationships, with the aim of bridging over 300 pathway databases. The Systems Biology ontology[13] targets the specific aspects of systems biology, especially in the context of computational modeling. The Foundational Model of Anatomy[14] is a domain ontology that represents a coherent body of explicit declarative knowledge about human anatomy. The Ontology for Physics in Biology[15] is also being developed, with specific reference to multiscale modeling.

Whereas the need for semantic standardization for the inputs and outputs of models is clear, in order to make them reusable, it is less clear where could be the need for ontologies of models themselves; this is an active research field for various groups. It should be noted that if we accept that models are published as web services,

[8] www.biomodels.net/sed-ml/
[9] www.oasis-open.org/committees/tc_home.php?wg_abbrev=wsbpel
[10] www.taverna.org.uk/
[11] www.gencontology.org/
[12] www.biopax.org/
[13] www.ebi.ac.uk/sbo
[14] http://fma.biostr.washington.edu/
[15] http://bioportal.bioontology.org/ontologies/38990

then this problem would turn into that of building semantic web services, a problem that is being extensively elaborated in the domain of knowledge management.

With respect to full simulations, the only concrete effort I am aware of is that of EBI, which is developing the Kinetic Simulation Algorithm Ontology[16] to annotate simulation descriptions encoded with SED-ML. Again, it remains unclear if or how standardized semantics would help, although in this case one can imagine future scenarios where a simulation involves dozens of models chosen from a library of thousands, in which case a standardized semantic would simplify the search and retrieval of the necessary components.

8.4.5   Sharing models

Let us for a moment assume that the problem previously posed has been solved, and that everyone can now transform the developed model into reusable components. The next logical step would be to share these artifacts with the rest of the research community. Data sharing is very popular in biological research with EMBL-Bank (DNA and RNA sequences), Ensembl (genomes), ArrayExpress (microarray-based gene-expression data), UniProt (protein sequences), InterPro (protein families, domains, and motifs) and PDBe (macromolecular structures), IntAct (protein–protein interactions), Reactome (pathways), and ChEBI (small molecules), among others. Much less is available with respect to other types of biomedical data, but some services are now beginning, such as PhysiomeSpace (Testi *et al.*, 2010).[17] For the models, some initiatives such as CellML or SBML operate repositories for models encoded with their standards.

A first objection to the idea of model sharing is that models cannot be shared, because outside the hands of their developers they are useless. My reaction to these objections is that when a model is useful only when used by one researcher, it is probably useless. If it is true that models capture tentative knowledge, sharing models is pretty much like sharing knowledge, a foundational concept of modern science.

The second group of objections is motivational in nature. In a complex modern world where the differences between scientists, policy makers, and entrepreneurs are becoming increasingly blurred, a researcher is not motivated to share with others an artifact that could have a commercial value, or that would give a competitive advantage to the researchers of another country in conflict with the researcher's own. But these problems are only apparently new; scientists have been facing them since the birth of science itself. Human beings can investigate nature for many reasons, but scientists do it only to increase the knowledge of the human species as a whole. Science is such only when we share with any other the tentative knowledge that we have developed. Anyone is welcome to become an entrepreneur or a scientific consultant for some governmental-sensitive projects, such as those

---

[16] www.ebi.ac.uk/compneur-srv/kisao/
[17] www.physiomespace.com/

related to defense, but when the circulation of knowledge is somehow limited, this is not science anymore.

The third and last group of objections to model sharing is pragmatic in nature. They revolve around the complexity and the time-consuming nature of transforming a dataset, a model, or a whole simulation into something useful for others, which can be shared effectively. There is a whole array of activities, such as re-implementation of our digital resources according to open standards, curation (i.e. annotation of the resource according to some established ontologies) that are long, tedious, and perceived as non-pertinent to our role as researchers. A first aspect is related to the quality of the instruments available for these activities, which are still very primitive, and frequently require a significant human effort. As these improve, the effort required to share a resource should be considerably decreased. But a more fundamental aspect is the consensus, still lacking, that modeling papers to be published should include the model itself as supplementary material. Many agree that this is the only way to ensure the reproducibility of numerical experiments, and are pushing journals to make this a requirement for publication. Needless to say, there is also some resistance both from authors and publishers, who see this as a considerable complication to the publishing process.

8.5 Creating truly integrative models

8.5.1 What is an integrative model?

In Chapter 1, I remarked how models can be distinguished as *descriptive* or *predictive*. In addition, predictive models can be further subdivided depending on the predominant type of logic reasoning used to build their idealizations: *inductive*, *deductive*, and *abductive* models.

Deductive models are models where the idealization is made primarily via deduction. To this family belong all physics-based mechanistic models. In these models we idealize the observed phenomenon according to a given set of physical laws, capable of capturing the aspects of the phenomenon that the model scope regards as relevant. All of the rest of the model is deduced from these general laws.

Inductive models are models where the idealization is made primarily via induction. To this category belong all models where the prediction is generated by induction, i.e. models based on the interpolation or extrapolation of observational data or measurements.

Abductive models are models where the idealization is made primarily via an abduction cycle. This is the most difficult but also the most interesting family of models. What are the models that change as we add new observations? So far three examples have been identified:

- *Neural networks* are models that change according to the training set with which we provide them, i.e. new observations;

- *Bayesian models* are models that change the probability associated with the model as we add new observations;
- *Integrative models* are models made of component idealizations and of relational idealizations defining the relations between components.

Integrative models are abductive: by adding a new component idealization *a* into our model, it predicts a surprising fact *b*; if we observe something close to *b* under the same conditions, the compositional idealization that forms the model is the fact *a*, which by abduction is suspected to be true.

This last category in particular is central to the discourse of this book. Indeed, multiscale models can be designed in two ways: one where a single model includes all the equations that describe the phenomenon at various space–time scales and the relationships that link them (typically using homogenization equations); the second as a collection of reductionist component models that are combined to form a model of models, a super-model, a hypermodel, to use just some of the names that have been used so far to indicate such constructs. I prefer the term *integrative models*, as defined previously.

8.5.2   Requirements of reusable integrative models

A multiscale model can thus be implemented as an integrative model, intended as the functional composition of component models that describe the phenomenon at different space–time scales, and of relation models, that define how these scale-specific processes interact with each other. In this sense, the construction of an integrative model can be seen much like the definition of a simulation, as suggested in Section 8.4. We need instruments that make it possible to select the appropriate models for each component or related element, and combine them so that they execute as a whole, with data flowing from one element to the other. The strategy being pursued by the advocates of neutral mark-up languages is that such composition can be done at the source-code level. If all the models that compose an integrative model are properly annotated according to some standardized semantics, so that each input and each output have standard names, in principle it should be possible to compose models stored in CellML or SBML and generate a new mark-up source code that describes the integrative model. Those who claim that models should be shared as services see the development of integrative models pretty much like the choreography of web services, or a problem of data flow management across remote procedures. In this sense, an integrative model would be a BPEL script or a Taverna script, which defines how various scale-specific models are combined to form the multiscale simulation.

A second set of requirements emerges from the execution constraints. As we move from deterministic integrative models made of a few component models to probabilistic or stochastic integrative models made of hundreds of component models, we need to ensure that the technology we use to build and run such integrative models can scale up accordingly. The approach based on monolithic models generated

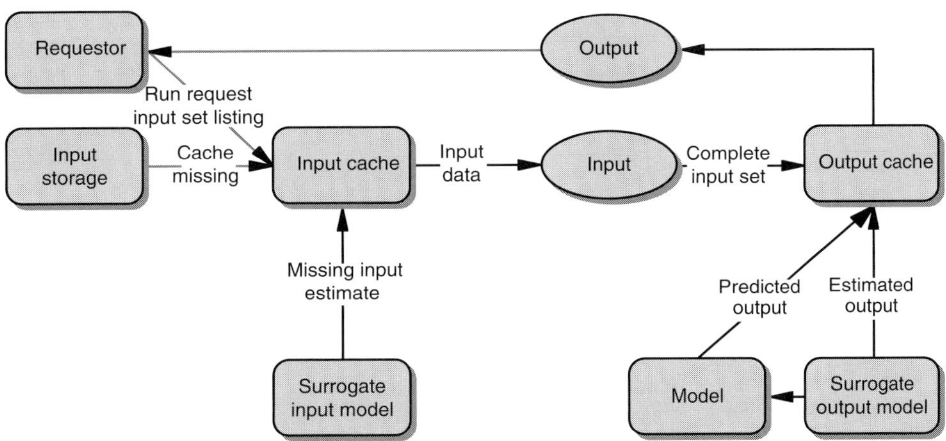

Figure 8.2   A possible scheme for a reusable component model, part of a larger RPC-based integrative model. The requestor submits a run request with the listing of the input set. If any element of the input set is not present in the cache, it is transferred from the input storage. If the requestor input set is incomplete, the surrogate input model can provide an estimate of the missing input elements. The complete input is then passed to the output cache; if a solution for that input set is already present, it is simply returned to the requestor. Otherwise, the surrogate output model rapidly generates an estimate of the output set; the model is run more slowly to produce the real predicted output set that is stored in the output cache.

by mashing-up the mark-up of the various component models suggests an execution environment limited to a single computational resource, be this a single computer, a cluster, a grid, etc. Such a condition is efficient, with minimal latency due to data transfer (in principle, in shared-memory architectures such a transfer could even happen in memory) but with serious problems of scalability.

The scenario of an integrative model described as the flow of data from one service to the other poses no problems of scalability, but quite a few of latency, as the data are passed from one service to the other. In this case, we should probably build within each model or service an *input cache*, a mechanism that makes it possible to avoid re-transferring large datasets that include the input set that was already transferred before; a typical example is when we run multiple simulations over the same 3D image dataset, such as a CT scan, but with different boundary conditions: the CT scan is the biggest part of the input set to be transferred, whereas boundary conditions can be a few bytes. Similarly, we should include an *output cache* to speed up requests for input sets that were already run, so reducing the model execution time (Figure 8.2).

Sometimes the input set is not identical to any input stored in the cache, but similar to some of them. In these cases, we can use *surrogate modeling* to generate an estimate of model's prediction rapidly. Surrogate modeling (Queipo *et al.*, 2005) is a cluster of methods for the generation of compact scalable analytic models that approximate the multivariate input–output behavior of complex systems, based on a limited set of computational expensive simulations. Thus, this approach can also

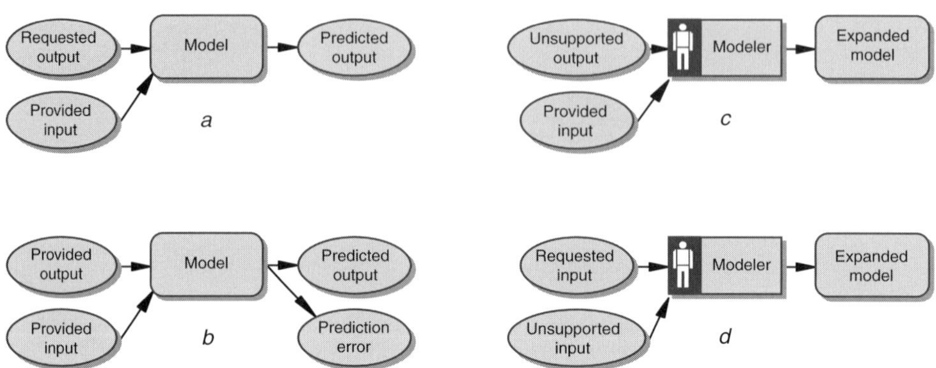

Figure 8.3   (a) When a complete input set is provided, and the model output is requested, the model runs in prediction mode. (b) When both input and output are provided the model runs in benchmark mode, and returns both the predicted output and the prediction error, intended as the difference between the predicted and provided outputs. (c, d) When the request contains an output or an input that is not supported by the current model implementation, the request is logged to the modeler, which can consider revising the model so as to extend the explanatory horizon in the direction requested.

be used to generate an estimate of the model output for a given input rapidly, once we have stored in the *output cache* enough input–output set pairs to identify the surrogate model properly. Some of the best surrogate modeling methods can also provide an estimate of the approximation error, i.e. of the difference between the estimated solution and the real one. In such cases, one can always generate a surrogate solution, leaving it to the requestor to decide whether the uncertainty is acceptable, or if a full run of the real model is required.

Another implementation problem is related to the fact that in many practical applications involving complex integrative models, the requestor (who requests a run of the model) does not have all necessary inputs required to identify the model fully. In these cases, the presence of a persistent cache for previous inputs can serve as a database to run a surrogate modeling of the input (whose most trivial instance is the average of all previous input instances). The surrogate inputs can then be used to complete the input set required by the model.

8.5.3   Supporting the inquiry cycle

Previously, I wrote that integrative modeling should be seen as a continuous abduction cycle, where the comparison of the predictions made by the actual version of the model with the available observations produces new and surprising facts that drive us toward the expansion of the explanatory horizon of the integrative model.

In principle, we should be able to conceive our integrative modeling technology so that it supports such a process, at least to some extent (Figure 8.3).

So, for example, if we expose our integrative model of bone remodeling, we might request a prediction for a given abnormal level of dietary calcium intake (i.e.

dietary calcium deficiency). If our model does not include the metabolic aspect, the request is rejected, but it is also logged to the model developer, who might consider further expanding it so as also to include this metabolic aspect.

As a spin-off, we can also consider the special case when the requestor provides both the input and the output. In such cases, the model should run in benchmark mode, make the prediction on the basis of the input, and then compare the predicted output with the output provided by the requestor, returning a prediction error (under the assumption that the provided output is "true"). If we keep track of all past runs in the various caches, this might become useful when we modify the model for some reason, as it would automatically re-run all previous simulations, including all provided benchmarks, telling us immediately if the changes we made actually improve the predictive accuracy with respect to the previously supported input–output sets.

8.6 Falsifying integrative models

8.6.1 Rationale

The problems posed by integrative models are not only technological in nature. As I tried to explain both in Chapter 1 and in this chapter, integrative models represent a whole new kind of model, substantially different from everything used previously in biomedical research. As such, there are some fundamental questions that are not usually considered when dealing with more conventional modeling approaches, which need to be revisited here, the most important being: how can we attempt to falsify an integrative model?

8.6.2 Once again: what is a model?

This book begins with this question; now after all these pages, it seems fair to pose it again. In Chapter 1, I defined modeling as a finalized cognitive activity. If we focus on scientific modeling, the scope of modeling should be that of science, or better that of the scientific method, to achieve "knowledge of reality that is object-ive, reliable, verifiable, and shareable." Thus, we can define scientific models as:

Models are finalized cognitive constructs of finite complexity that idealize an infinitely complex portion of reality through idealizations that contribute to the achievement of a knowledge on that portion of reality that is objective, shareable, reliable, and verifiable.

- *Objective* is the opposite of subjective: agreeable by many peers, mostly unbiased by personal views. In empirical sciences, no model will ever be entirely objective, this is an abstract concept, unachievable in full, but to which nevertheless we all should aim. If we recognize that the search for scientific truth is a collective endeavor, then the search for objectivity is strongly linked with the second attribute, shareability.

- *Shareable* means that it can be shared with our peers. Again, with a prescriptive perspective objectivity and shareability are strongly related with the need for reproducibility.
- *Reliable* means "that can be trusted." In this context, we can define this property as the ability of a model to provide comparable quality of performance within a given range of conditions. What the quality of performance is depends on the finality of the model, which must therefore always be stated explicitly. We also recognize that almost no model remains reliable for every possible range of conditions (or more correctly for every possible state that the portion of reality under examination can assume). This range is defined as the limit of validity of the model, and it must also be stated explicitly each time we develop a scientific model. The limit of validity should not be confused with the explanatory horizon introduced earlier in this chapter, although the two are usually closely related.
- *Verifiable* means "conceived and presented so that the quality of its performance can be assessed."

While the first three attributes are perfectly in accordance with the discussion on previous pages, the attribute of verifiability requires some additional thought, as it apparently contradicts Popper's position on falsifiability.

8.6.3   Falsification vs. verification

In his monumental book, *The Logic of Scientific Discovery* (Popper, 1992) Karl Popper introduced the most important pillar of the cathedral of thought that is his philosophy of science: the asymmetry between falsification and verification. Popper starts from the recognition that all scientific theories are irreducibly hypothetical and convincingly argues that while it would take an infinite number of positive observations to confirm a theory conclusively, a single piece of counter-evidence is sufficient to prove it false. Thus, theories can only be proved false, never confirmed conclusively. Indeed, Popper proposes that the best way to "measure" how genuinely scientific a theory is its *falsifiability*, i.e. whether it is formulated in ways that make it logically (but also practically) possible to disprove it.

This position apparently seems to contradict the idea that a predictive model can be validated. To resolve this apparent contradiction we must look closer at the human activity we call science. Popper suggests that human beings are pushed to produce new knowledge in order to solve problems that rise in the specific socio-historical context. But this definition, while agreeable, tends to confuse problem-solving and knowledge-seeking activities. If we think in terms of predictive models, a problem can be solved pretty well with a model that is accurate in its predictions. However, such a model does not necessarily provide an explanation of why the phenomenon being predicted happens.

A model in itself can never be true or false. At most we can validate it, in the sense of measuring its predictive accuracy with respect to the outcomes of one or more controlled experiments. If we limit our goal to the prediction, to the solution of the problem, this can be considered enough. Of course nothing can guarantee that in different conditions the model will retain that predictive accuracy we tested for a finite set of cases. This is why in physics-based disciplines, such as engineering, while we use non-explanatory and purely phenomenological models whenever necessary, we prefer to resort to mechanistic models whenever possible.

The difference between an explanatory and a non-explanatory model is that the former contains a theory, a candidate explanation on why the thing we are trying to predict actually happens. That theoretical element however, cannot be confirmed, but only falsified. So while the *predictions of a model can be validated*, the *theory that informs an explanatory model can only be falsified*, and at most we can provide in its support a long list of unsuccessful attempts of falsification.

8.6.4 Idealization and truth content

Having clarified what we mean by validation, let us now look at the falsification of mechanistic models. One possible approach to developing generic strategies for falsification is to look at the predominant type of logical reasoning used to formulate the idealizations that form the model. As already discussed, there are three recognized types of logical reasoning:

- *Deduction*: deriving a as a consequence of b;
- *Induction*: inferring a from multiple instantiations of b;
- *Abduction*: inferring a as an explanation of b.

8.6.4.1 Falsification of deductive models

To this family belong all physics-based mechanistic models. In these models we idealize the observed phenomenon according to a given set of physical laws, capable of capturing the aspects of the phenomenon that the model's scope regards as relevant. All the rest of the model is deduced from these general laws. Typical examples are solid- and fluid-mechanical models used in biomechanics (e.g. (Schileo *et al.*, 2007; Singh *et al.*, 2010)) or the electro-chemo-mechanical models used in cardiac physiology (see, for example, Bassingthwaighte's review article on the cardiac physiome (Bassingthwaighte *et al.*, 2009)).

The potential truth content of deduction is totally dependent on the truth content of the initial assumptions: if the precedents are true, deduction also ensures that the consequents are true. If the laws we choose to represent the phenomenon of interest are (i) scientifically true and (ii) accurate in capturing the relevant aspects of the phenomenon of interest, everything deduced from them has the same truth content.

Another important property of deductive models is potential generality. If the laws I used as antecedents are sufficiently general, I might find that the same model is effective in predicting a very large set of similar phenomena.

Because in a deductive model the truth content of the consequents derives from that of the inputs, deductive models are usually challenged by questioning the laws of physics or of biology that we used to build our deductive model. As these laws have frequently already been challenged for a long time, the point is not to prove the laws false in themselves, but rather to challenge the assumption that the phenomenon under observation can be accurately modeled in its aspects relevant to the modeling scope by that choice of laws. So, for example, if we idealize the biomechanical behavior of the mineralized extracellular matrix of bones with the theory of elasticity, to falsify this idealization we should conduct experiments to see if, within the range of physiological skeletal loading, the bone tissue shows a significant deviation from the proportionality in the relationship between stresses and strain. This does not involve attempting the falsification of the entire theory of elasticity, but simply its application to bone tissue subjected to certain loading conditions, temperature, etc.

The other falsification strategy is to investigate phenomena that while independent from the one modeled logically descend from the idealization made. So to stick with the bone example, if bone behavior in physiological conditions is linearly elastic, then in the range of deformation rates that is physiologically possible bones should exhibit no significant viscous dissipation, which can be verified with isothermal experiments (Yamashita *et al.*, 2002).

8.6.4.2   Falsification of inductive models

To this category belong all models where the prediction is generated by induction, i.e. models based on the interpolation or extrapolation of observational data or measurements. Typical examples are predictive models based on epidemiological studies (e.g. population-based models to predict the risk of bone fractures in osteoporotic patients (Gauthier *et al.*, 2010)).

In models where the idealizations are predominantly formulated by induction, within a deterministic framework the potential truth content is null. In fact, if an inductive model accurately predicts a phenomenon n times, nothing guarantees that for the $(n + 1)$th set of parameters the model will also be accurate. In a probabilistic framework, however, which postulates that all physical phenomena show a certain degree of regularity, the potential truth content of an inductive model depends on statistical indicators, such as level of significance, power of the test, etc.

The first approach to the falsification of inductive models should also be inductive. We frequently forget that if we want to use data regression to make predictions, it is not sufficient to verify that the regression is accurate "on average." In this sense, attempts to falsify an inductive model by verifying whether the predicted values and the measured values simply correlate should be considered largely inadequate. Of course, predictive models must make predictions that correlate with the observations they intend to model; but predicting is much more than correlating; what we must do is to verify whether the predictive error remains within the acceptance limits over the entire range of validity of the model, for every prediction.

Another way of looking at this problem is to recognize that the peak predictive error that we observe over a large set of validation experiments should be considered as the interval of uncertainty of our model's predictions. If we run our model with two different sets of parameters, and the two predictions differ by less than the peak error we observed during the validation experiments, nothing can be concluded, because we cannot positively confirm that the difference really exists and is not due to the inadequacy of our model.

But this is not sufficient. Inductive models must also be challenged on the statistical side. If the model assumes that the data are normally distributed, are they? What is the level of statistical significance for our predictor? What is the statistical power? These tests should be made both on the data that we use to inform our inductive model and also on the predictions produced by our model. A typical example is the following: say that we want to predict the surface principal strains in a long bone subjected to some bending. This means that on one side of the bone strains will be largely positive, on the opposite side largely negative, and in the two other sides close to zero, as they are close to the neutral axis. Now if we plot our predictions against some experimental measurements taken with strain gages, even if the model is quite wrong, we shall always obtain pretty high coefficients of regression. This is because the data are polarized; even if our data distribution comprised two circles, one in the positive–positive quadrant and one in the negative–negative quadrant, the global regression coefficient would be pretty high. But if we were to make separate regressions, one for the positive and one for the negative, we would find nearly zero regression coefficients. In reality if we do separate regression analysis for negative and positive values, we might find that the two resulting regressions are significantly different in slope or intercept from the one obtained by pooling all data together. This is an example of a regression that is statistically significant, but that has a very low statistical power.

8.6.4.3 Abductive models

This is the most difficult but also the most interesting family of models. What are the models that change as we add new observations? So far, I have identified three examples:

- *Neural networks* are models that change according to the training set with which we provide them, i.e. new observations;
- *Bayesian models* are models that change the probability associated with models as we add new observations;
- *Integrative models* are models made of component idealizations and of relational idealizations defining the relations between components.

This last category in particular is central to the scope of this book. However, at present, very little can be said about the potential truth content of abductive models, and the best strategies to falsify them. The best advice for integrative models is first to attempt the falsification of each component idealization and of each relational idealization separately, and then repeat it for the integrative model considered as a whole.

However, this strategy does not apply to Bayesian models or to neural networks; thus a more general falsification framework, valid for all abductive models, needs to be expounded. A possible direction of research is that of the so-called bootstrap methods (Wu, 1986): repeating the construction of the model through the abduction cycle using multiple sets of independent observations all relative to the phenomenon to be modeled, should yield the same model.

8.7   Conclusions

In this last chapter, I have tried to present multiscale modeling, and in particular the specific approach to multiscale modeling called integrative modeling, from a methodological and technological point of view.

What clearly emerges from this description is that, at present, integrative models can only be built manually, with no technological support, which considerably limits the size and complexity that we can address. However, the research community is working hard to develop the Virtual Physiological Human, a framework of methods and technologies that, once developed, will enable the integrative investigation of human physiopathology. In this chapter, I summarize where such research efforts lie, the preliminary results, and the grand challenges that still lie ahead.

Not everything described in this section exists, or exists only as a truly preliminary and incomplete prototype. Thus, no one has any idea of how these concepts will turn out when brought into practice. This is a whole uncharted territory, which lies on the boundary between biomedicine, mathematics, computer science, biomedical engineering, and physics. But the scenarios described here would radically transform biomedical simulation, and possibly simulation at large; in this sense, undertaking such challenging developments seems worth the risk.

The problem of identifying integrative models is probably the one being most aggressively addressed in current research. We are optimistic that this will become less and less a problem, especially if the first clinical trials will prove the efficacy of simulation-based medical technology, orienting all medical instrumentation toward the production of quantitative multiscale information.

The ability to provide an effective support for the inquiry cycle with the modeling technology is a promising horizon, a research topic that has been left almost untouched so far.

The transformation of models into reusable quanta of knowledge implies the development of complex technologies and methods, and there is still some controversy on the best approach. Certainly, the standardization of minimum-requirement specifications, on data, modeling, and simulation syntaxes and semantics, as well as on the mechanisms for sharing models is essential to this goal. But it might be necessary to go much beyond that, developing specific technologies that package each model and transform it into a public service complete with complex caching and logging services.

The last point, touching the discussion on the falsifiability of integrative models, is more complex; it is tempting to say that their recursive nature perfectly matches the process of scientific research, and because of this they promise to embody the scientific method better than any other modeling approach. However, the current incompleteness of the falsification framework for this family of models casts some concerns on their use. Because of this, I strongly recommend that the scientific community, possibly with the help of the philosophy of science community, adopt the problem of the falsification of abductive models as a central goal for future research.

References

Ackerman, M. J. (1998). The Visible Human Project: a resource for anatomical visualization. *Stud. Health Technol. Inform.*, **52**(2), 1030–2.

Adams, M. F., Bayraktar, H. H., Keaveny, T. M., and Papadopoulos, P. (2004). Ultrascalable implicit finite element analyses in solid mechanics with over a half a billion degrees of freedom. In *Proceedings of the 2004 ACM/IEEE Conference on Supercomputing.*

Albracht, K., Arampatzis, A., and Baltzopoulos, V. (2008). Assessment of muscle volume and physiological cross-sectional area of the human triceps surae muscle in vivo. *J. Biomech.*, **41**(10), 2211–18.

An, K. N., Takahashi, K., Harrigan, T. P., and Chao, E. Y. (1984). Determination of muscle orientations and moment arms. *J. Biomech. Eng.*, **106**(3), 280–2.

Andersen, T. L., Sondergaard, T. E., Skorzynska, K. E., *et al.* (2009). A physical mechanism for coupling bone resorption and formation in adult human bone. *Am. J. Pathol.*, **174**(1), 239–47.

Andersen, T. L., Soe, K., Sondergaard, T. E., Plesner, T., and Delaisse, J. M. (2010). Myeloma cell-induced disruption of bone remodelling compartments leads to osteolytic lesions and generation of osteoclast-myeloma hybrid cells. *Br. J. Haematol.*, **148**(4), 551–61.

Ardizzoni, A. (2001). Osteocyte lacunar size – lamellar thickness relationships in human secondary osteons. *Bone*, **28**(2), 215–19.

Aristotle. (350 BC) *Metaphysics.* **Book H**(1045b), 8–10.

Ascher, U. M. and Petzold, L. R. (1998). *Computer Methods for Ordinary Differential Equations and Differential-Algebraic Equations.* SIAM.

Baca, V., Horak, Z., Mikulenka, P., and Dzupa, V. (2008). Comparison of an inhomogeneous orthotropic and isotropic material models used for FE analyses. *Med. Eng. Phys.*, **30**(7), 924–30.

Bassingthwaighte, J. B. (2000). Strategies for the physiome project. *Ann. Biomed. Eng.*, **28**(8), 1043–58.

Bassingthwaighte, J., Hunter, P., and Noble D. (2009). The Cardiac Physiome: perspectives for the future. *Exp. Physiol.*, **94**(5), 597–605.

Bathe, K. J. and Wilson, E. L. (1976). *Numerical Methods in Finite Element Analysis.* Prentice-Hall.

Behari, J. (2009). Elements of bone biophysics. In *Biophysical Bone Behavior: Principles and Applications.* John Wiley & Sons.

Bekas, C., Curioni, A., Arbenz, P., *et al.* (2008). Extreme scalability challenges in micro-finite element simulations of human bone. In *International Supercomputing Conference.* ISC.

Benameur, S., Mignotte, M., Labelle, H., and De Guise, J. A. (2005). A hierarchical statistical modeling approach for the unsupervised 3-D biplanar reconstruction of the scoliotic spine. *IEEE Trans. Biomed. Eng.*, **52**(12), 2041–57.

Bensamoun, S. F., Ringleb, S. I., Chen, Q., *et al.* (2007). Thigh muscle stiffness assessed with magnetic resonance elastography in hyperthyroid patients before and after medical treatment. *J. Magn. Reson. Imaging*, **26**(3), 708–13.

Beraudi, A., Stea, S., Bordini, B., Baleani, M., and Viceconti, M. (2010). Osteon classification in human fibular shaft by circularly polarized light. *Cells Tissues Organs*, **191**(3), 260–8.

Bergmann, G., Deuretzbacher, G., Heller, M., *et al.* (2001). Hip contact forces and gait patterns from routine activities. *J. Biomech.*, **34**(7), 859–71.

Bernstein, N. (1967). *The Co-ordination and Regulation of Movements.* Pergamon Press.

Bevill, G., Farhamand, F., and Keaveny, T. M. (2009). Heterogeneity of yield strain in low-density versus high-density human trabecular bone. *J. Biomech.*, **42**(13), 2165–70.

Bikle, D. D., Sakata, T., and Halloran, B. P. (2003). The impact of skeletal unloading on bone formation. *Gravit. Space Biol. Bull.*, **16**(2), 45–54.

Bilezikian, J. P., Raisz, L. G., and Martin, T. J., eds. (2008). *Principles of Bone Biology.* 3rd edn. Academic Press.

Blemker, S. S., Pinsky, P. M., and Delp, S. L. (2005). A 3D model of muscle reveals the causes of nonuniform strains in the biceps brachii. *J. Biomech.*, **38**(4), 657–65.

Boutroy, S., Bouxsein, M. L., Munoz, F., and Delmas, P. D. (2005). In vivo assessment of trabecular bone microarchitecture by high-resolution peripheral quantitative computed tomography. *J. Clin. Endocrinol. Metab.*, **90**(12), 6508–15.

Brower, A. C. and Allman, R. M. (1981). Pathogenesis of the neurotrophic joint: neurotraumatic vs. neurovascular. *Radiology*, **139**(2), 349–54.

Brown, T. D. and Ferguson, A. B., Jr. (1980). Mechanical property distributions in the cancellous bone of the human proximal femur. *Acta Orthop. Scand.*, **51**(3), 429–37.

Buchanan, T. S., Lloyd, D. G., Manal, K., and Besier, T. F. (2004). Neuromusculoskeletal modeling: estimation of muscle forces and joint moments and movements from measurements of neural command. *J. Appl. Biomech.*, **20**(4), 367–95.

Burr, D. B., Robling, A. G., and Turner, C. H. (2002). Effects of biomechanical stress on bones in animals. *Bone*, **30**(5), 781–6.

Calvo, B., Ramirez, A., Alonso, A., *et al.* (2010). Passive nonlinear elastic behaviour of skeletal muscle: experimental results and model formulation. *J. Biomech.*, **43**(2), 318–25.

Campanacci, M. (1999). *Bone and Soft Tissue Tumors: Clinical Features, Imaging, Pathology and Treatment.* Springer Verlag.

Campbell, D. T. (1974). Downward causation in hierarchically organized biological systems. In *Studies in the Philosophy of Biology and Related Problems*, ed. P. J. Ayala and T. Dobzhansky. University of California Press, pp. 179–86.

Cappozzo, A., Della Croce, U., Leardini, A., and Chiari, L. (2005). Human movement analysis using stereophotogrammetry. Part 1: theoretical background. *Gait Posture*, **21**(2), 186–96.

Carins, J. (2005). *Nutritional Determinants of Bone Health: A Survey of Australian Defence Force (ADF) Trainees.* http://dspace.dsto.defence.gov.au/dspace/handle/1947/4320. Accessed May 7th, 2010.

Carnap, R. (1958). *Introduction to Symbolic Logic and Its Applications.* Dover Publications.

Carter, D. R. and Hayes, W. C. (1977). The compressive behavior of bone as a two-phase porous structure. *J. Bone Joint Surg. Am.*, **59**(7), 954–62.

Casas-Ganem, J. and Healey, J. H. (2005). Advances that are changing the diagnosis and treatment of malignant bone tumors. *Curr. Opin. Rheumatol.*, **17**(1), 79–85.

Chaibi, Y., Cresson, T., Aubert, B., *et al.* (2011), Fast 3D reconstruction of the lower limb using a parametric model and statistical inferences and clinical measurements calculation from biplanar X-rays. *Comput. Methods Biomech. Biomed. Engin.*, **14**(1).

Chang, P. B., Williams, B. J., Bhalla, K. S., *et al.* (2001). Design and analysis of robust total joint replacements: finite element model experiments with environmental variables. *J. Biomech. Eng.*, **123**(3), 239–46.

Chau, J. F., Leong, W. F., and Li, B. (2009). Signaling pathways governing osteoblast proliferation, differentiation and function. *Histol. Histopathol.*, **24**(12), 1593–606.

Chenu, C. and Marenzana, M. (2005). Sympathetic nervous system and bone remodeling. *Joint Bone Spine*, **72**(6), 481–3.

Clancy, E. A., Morin, E. L., and Merletti, R. (2002). Sampling, noise-reduction and amplitude estimation issues in surface electromyography. *J. Electromyogr. Kinesiol.*, **12**(1), 1–16.

Clapworthy, G., Kohl, P., Gregerson, H., *et al.* (2007). Digital human modelling: a global vision and a European perspective. In *Digital Human Modeling*. Springer-Verlag, pp. 549–58.

Clapworthy, G., Viceconti, M., Coveney, P. V., and Kohl, P. (2008). The virtual physiological human: building a framework for computational biomedicine I. Editorial. *Philos. Transact. A Math. Phys. Eng. Sci.*, **366**(1878), 2975–8.

Collins, J. J. (1995). The redundant nature of locomotor optimization laws. *J. Biomech.*, **28**(3), 251–67.

Confavreux, C. B., Levine, R. L., and Karsenty, G. (2009). A paradigm of integrative physiology, the crosstalk between bone and energy metabolisms. *Mol. Cell. Endocrinol.*, **310**(1–2), 21–9.

Cook, R. D. (1995). *Finite Element Modeling for Stress Analysis*. John Wiley & Sons.

Cooper, D. M., Turinsky, A. L., Sensen, C. W., and Hallgrimsson, B. (2003). Quantitative 3D analysis of the canal network in cortical bone by micro-computed tomography. *Anat. Rec. B New Anat.*, **274**(1), 169–79.

Cowin, S. C. (1988). *Bone Mechanics*. CRC.

Cowin, S. C. and Hegedus, D. H. (1976). Bone remodeling I: theory of adaptive elasticity. *J. Elast.*, **6**(3), 313–26.

Cristofolini, L. (1997). A critical analysis of stress shielding evaluation of hip prostheses. *Crit. Rev. Biomed. Eng.*, **25**(4–5), 409–83.

Cristofolini, L., Cappello, A., and Toni, A. (1994). Experimental errors in the application of photoelastic coatings on human femurs with uncemented hip stems. *Strain*, **30**(3), 95–104.

Cristofolini, L., Taddei, F., Baleani, M., *et al.* (2008). Multiscale investigation of the functional properties of the human femur. *Philos. Transact. A Math. Phys. Eng. Sci.*, **366**(1879), 3319–41.

Damm, P., Graichen, F., Rohlmann, A., Bender, A., and Bergmann, G. (2010). Total hip joint prosthesis for in vivo measurement of forces and moments. *Med. Eng. Phys.*, **32**(1), 95–100.

Daumer, M., Thaler, K., Kruis, E., *et al.* (2007). Steps towards a miniaturized, robust and autonomous measurement device for the long-term monitoring of patient activity: ActiBelt. *Biomed. Tech. (Berl.)*, **52**(1), 149–55.

Dawkins, R. (1976). *The Selfish Gene*. Oxford University Press.

Drillis, R., Contini, R., Bluestein, M. (1964). Body segment parameters; a survey of measurement techniques. *Artif. Limbs*, **25**, 44–66.

Durkin, J. L. and Dowling, J. J. (2003). Analysis of body segment parameter differences between four human populations and the estimation errors of four popular mathematical models. *J. Biomech. Eng.*, **125**(4), 515–22.

Durkin, J. L., Dowling, J. J., and Andrews, D. M. (2002). The measurement of body segment inertial parameters using dual energy X-ray absorptiometry. *J. Biomech.*, **35**(12), 1575–80.

Edidin, A. A. (1991). *Modelling the distribution of bone moduli in bone-implant systems.* Ph.D. thesis. Ithaca: Cornell University, New York.

Edman, K. A. (1979). The velocity of unloaded shortening and its relation to sarcomere length and isometric force in vertebrate muscle fibres. *J. Physiol.*, **291**, 143–59.

Eich-Soellner, E. and Führer C. (1998). *Numerical Methods in Multibody Dynamics.* Teubner.

Elefteriou, F. (2008). Regulation of bone remodeling by the central and peripheral nervous system. *Arch. Biochem. Biophys.*, **473**(2), 231–6.

Eppell, S. J., Smith, B. N., Kahn, H., and Ballarini, R. (2006). Nano measurements with micro-devices: mechanical properties of hydrated collagen fibrils. *J. R. Soc. Interface*, **3**(6), 117–21.

Erdemir, A., McLean, S., Herzog, W., and van den Bogert, A. J. (2007). Model-based estimation of muscle forces exerted during movements. *Clin. Biomech.*, **22**(2), 131–54.

Evans, F. G. (1957). *Stress and Strain in Bones. Their Relation to Fractures and Osteogenesis.* Charles C. Thomas.

Evans, F. G. (1973a). *Mechanical Properties of Bone.* Charles C. Thomas.

Evans, F. G. (1973b). Factors affecting the mechanical properties of bone. *Bull. N Y Acad. Med.*, **49**(9), 751–64.

Farina, D., Merletti, R., and Enoka, R. M. (2004). The extraction of neural strategies from the surface EMG. *J. Appl. Physiol.*, **96**(4), 1486–95.

Fenner, J. W., Brook, B., Clapworthy, G., *et al.* (2008). The EuroPhysiome, STEP and a roadmap for the virtual physiological human. *Philos. Transact. A Math. Phys. Eng. Sci.*, **366**(1878), 2979–99

Fernandez, J. W., Buist, M. L., Nickerson, D. P., and Hunter, P. J. (2005). Modelling the passive and nerve activated response of the rectus femoris muscle to a flexion loading: a finite element framework. *Med. Eng. Phys.*, **27**(10), 862–70.

Fratzl-Zelman, N., Roschger, P., Misof, B. M., *et al.* (2009). Normative data on mineralization density distribution in iliac bone biopsies of children, adolescents and young adults. *Bone*, **44**(6), 1043–8.

Frigg, R. and Hartmann, S. (2006). Models in Science. In *Stanford Encyclopedia of Philosophy*. http://plato.stanford.edu/archives/fall2008/entries/models-science/. Accessed May 10th, 2011.

Frost, H. M. (1997). On our age-related bone loss: insights from a new paradigm. *J. Bone. Miner. Res.*, **12**(10), 1539–46.

Fung, Y. C. (1993). *Biomechanics: Mechanical Properties of Living Tissues.* Springer.

Fung, Y. C. and Tong, P. (2001). *Classical and Computational Solid Mechanics.* World Scientific.

Ganss, B., Kim, R. H., and Sodek, J. (1999). Bone sialoprotein. *Crit. Rev. Oral Biol. Med.*, **10**(1), 79–98.

Gauthier, A., Kanis, J. A, Martin, M., *et al.* (2011). Development and validation of a disease model for postmenopausal osteoporosis. *Osteoporos. Int.*, **22**(3), 771–80.

Ghert, M., Colterjohn, N., and Manfrini, M. (2007). The use of free vascularized fibular grafts in skeletal reconstruction for bone tumors in children. *J. Am. Acad. Orthop. Surg.*, **15**(10), 577–87.

Giangregorio, L. and Blimkie, C. J. (2002). Skeletal adaptations to alterations in weight-bearing activity: a comparison of models of disuse osteoporosis. *Sports Med.*, **32**(7), 459–76.

Gilles, B. and Magnenat-Thalmann, N. (2010). Musculoskeletal MRI segmentation using multi-resolution simplex meshes with medial representations. *Med. Image Anal.*, **14**(3), 291–302.

Glisson, R. R., Musgrave, D. S., Graham, R. D., and Vail, T. P. (2000). Validity of photoelastic strain measurement on cadaveric proximal femora. *J. Biomech. Eng.*, **122**(4), 423–9.

Gordon, A. M., Huxley, A. F., and Julian, F. J. (1966). The variation in isometric tension with sarcomere length in vertebrate muscle fibres. *J. Physiol.*, **184**(1), 170–92.

Gordon, J. A., Tye, C. E., Sampaio, A. V., *et al.* (2007). Bone sialoprotein expression enhances osteoblast differentiation and matrix mineralization in vitro. *Bone*, **41**(3), 462–73.

Grassi, L., Hraiech, N., Schileo, E., *et al.* (2010). Evaluation of the generality and accuracy of a new mesh morphing procedure for the human femur. *Med. Eng. Phys.*, **33**(1), 112–20.

Grätsch, T. and Bathe, K. J. (2005). A posteriori error estimation techniques in practical finite element analysis. *Comput. Struct.*, **83**(4–5), 235–65.

Gray, H. (1918). *Gray's Anatomy: The Anatomical Basis of Medicine and Surgery*. 20th edn. Longman.

Harrigan, T. P., Jasty, M., Mann, R. W., and Harris, W. H. (1988). Limitations of the continuum assumption in cancellous bone. *J. Biomech.*, **21**(4), 269–75.

Hauge, E. M., Qvesel, D., Eriksen, E. F., Mosekilde, L., Melsen, F. (2001). Cancellous bone remodeling occurs in specialized compartments lined by cells expressing osteoblastic markers. *J. Bone Miner. Res.*, **16**(9), 1575–82.

Heinegard, D. and Oldberg, A. (1989). Structure and biology of cartilage and bone matrix noncollagenous macromolecules. *FASEB J.*, **3**(9), 2042–51.

Helgason, B., Taddei, F., Palsson, H., *et al.* (2008). A modified method for assigning material properties to FE models of bones. *Med. Eng. Phys.*, **30**(4), 444–53.

Homminga, J., Huiskes, R., Van Rietbergen, B., Ruegsegger, P. and Weinans, H. (2001). Introduction and evaluation of a gray-value voxel conversion technique. *J. Biomech.*, **34**(4), 513–17.

Horowitz, A. and Pollack G. H. (1993). Force–length relation of isometric sarcomeres in fixed-end tetani. *Am. J. Physiol.*, **264**(1 Pt 1), C19–26.

Hoyt, K., Kneezel, T., Castaneda, B., and Parker K. J. (2008). Quantitative sonoelastography for the in vivo assessment of skeletal muscle viscoelasticity. *Phys. Med. Biol.*, **53**(15), 4063–80.

Hsiao, H., Guan, J., and Weatherly, M. (2002). Accuracy and precision of two in-shoe pressure measurement systems. *Ergonomics*, **45**(8), 537–55.

Hunter, G. K. and Goldberg, H. A. (1993). Nucleation of hydroxyapatite by bone sialoprotein. *Proc. Natl. Acad. Sci. USA*, **90**(18), 8562–5.

Hunter, P. J. and Viceconti, M. (2009). The VPH-Physiome project: standards and tools for multiscale modeling in clinical applications. *IEEE Rev. Biomed. Eng.*, **2**, 40–53.

Hunter, P., Coveney, P. V., de Bono, B., *et al.* (2010). A vision and strategy for the virtual physiological human in 2010 and beyond. *Philos. Transact. A Math. Phys. Eng. Sci.*, **368**(1920), 2595–614.

Imai, K., Ohnishi, I., Bessho, M., and Nakamura, K. (2006). Nonlinear finite element model predicts vertebral bone strength and fracture site. *Spine*, **31**(16), 1789–94.

Imai, Y., Youn, M. Y., Kondoh, S., *et al.* (2009). Estrogens maintain bone mass by regulating expression of genes controlling function and life span in mature osteoclasts. *Ann. NY Acad. Sci.*, **1173**(Suppl 1), E31–9.

Innocenti, M., Abed, Y. Y., Beltrami, G., *et al.* (2009). Biological reconstruction after resection of bone tumors of the proximal tibia using allograft shell and intramedullary free vascularized fibular graft: long-term results. *Microsurgery*, **29**(5), 361–72.

Jedlicka, P. (2010). Ewing sarcoma, an enigmatic malignancy of likely progenitor cell origin, driven by transcription factor oncogenic fusions. *Int. J. Clin. Exp. Pathol.*, **3**(4), 338–47.

Jones, E. J., Bishop, P. A., Woods, A. K., and Green J. M. (2008). Cross-sectional area and muscular strength: a brief review. *Sports Med.*, **38**(12), 987–94.

Jung, C. G. (1981). *The Archetypes and the Collective Unconscious*. 2nd edn. Princeton University Press.

Kalender, W. A. (2005). *Computer Tomography*. Publicis Corporate Publishing.

Kang, Y., Engelke, K., and Kalender, W. A. (2003). A new accurate and precise 3-D segmentation method for skeletal structures in volumetric CT data. *IEEE Trans. Med. Imaging*, **22**(5), 586–98.

Kanis, J. A., Oden, A., Johnell, O., *et al.* (2007). The use of clinical risk factors enhances the performance of BMD in the prediction of hip and osteoporotic fractures in men and women. *Osteoporos Int.*, **18**(8), 1033–46.

Kannus, P., Sievanen, H., and Vuori, I. (1996). Physical loading, exercise, and bone. *Bone*, **18**(1 Suppl), 1S–3S.

Karsenty, G. (2001). Leptin controls bone formation through a hypothalamic relay. *Recent Prog. Horm. Res.*, **56**, 401–15.

Kavukcuoglu, N. B., Patterson-Buckendahl, P., and Mann, A. B. (2009). Effect of osteocalcin deficiency on the nanomechanics and chemistry of mouse bones. *J. Mech. Behav. Biomed. Mater.*, **2**(4), 348–54.

Keaveny, T. M., Pinilla, T. P., Crawford, R. P., Kopperdahl, D. L., and Lou, A. (1997). Systematic and random errors in compression testing of trabecular bone. *J. Orthop. Res.*, **15**(1), 101–10.

Keyak, J. H., Meagher, J. M., Skinner, H. B., and Mote, C. D., Jr. (1990). Automated three-dimensional finite element modelling of bone: a new method. *J. Biomed. Eng.*, **12**(5), 389–97.

Khosla, S. (2001). Minireview: the OPG/RANKL/RANK system. *Endocrinology*, **142**(12), 5050–5.

Kleissen, R. F., Buurke, J. H., Harlaar, J., and Zilvold, G. (1998). Electromyography in the biomechanical analysis of human movement and its clinical application. *Gait Posture*, **8**(2), 143–58.

Knothe Tate, M. L. (2003). "Whither flows the fluid in bone?" An osteocyte's perspective. *J. Biomech.*, **36**(10), 1409–24.

Kohl, P. and Viceconti, M. (2010). The virtual physiological human: computer simulation for integrative biomedicine II. *Philos. Transact. A Math. Phys. Eng. Sci.*, **368**(1921), 2837–9.

Kohl, P., Coveney, P., Clapworthy, G., and Viceconti, M. (2008). The virtual physiological human. Editorial. *Philos. Transact. A Math. Phys. Eng. Sci.*, **366**(1879), 3223–4.

Krokos, M., Podgorelec, D., Clapworthy, G. J., *et al.* (2005). Patient-specific muscle models for surgical planning. *Third International Conference on Medical Information Visualisation – Biomedical Visualisation (MediVis 2005)*. pp. 3–8.

Krum, S. A., Miranda-Carboni, G. A., Hauschka, P. V., *et al.* (2008). Estrogen protects bone by inducing Fas ligand in osteoblasts to regulate osteoclast survival. *EMBO J.*, **27**(3), 535–45.

Lattanzi, R., Baruffaldi, F., Zannoni, C., and Viceconti, M. (2004). Specialised CT scan protocols for 3-D pre-operative planning of total hip replacement. *Med. Eng. Phys.*, **26**(3), 237–45.

Laz, P. J., Stowe, J. Q., Baldwin, M. A., Petrella, A. J., and Rullkoetter, P. J. (2007). Incorporating uncertainty in mechanical properties for finite element-based evaluation of bone mechanics. *J. Biomech.*, **40**(13), 2831–6.

Leardini, A., Chiari, L., Della Croce, U., and Cappozzo, A. (2005). Human movement analysis using stereophotogrammetry. Part 3. Soft tissue artifact assessment and compensation. *Gait Posture*, **21**(2), 212–25.

Leardini, A., Taddei, F., Manfrini, M., Viceconti, M., and Benedetti, M. G. (2007). Gait analysis and bone model registration in patients operated on massive skeletal reconstruction: a weighted-joint technique for artifact reduction. *J. Biomech.*, **40**(Supplement 2), S553.

LeBlanc, A. D., Spector, E. R., Evans, H. J., and Sibonga, J. D. (2007). Skeletal responses to space flight and the bed rest analog: a review. *J. Musculoskelet. Neuronal Interact.*, **7**(1), 33–47.

Lee, Y. S., Seon, J. K., Shin, V. I., Kim, G. H., and Jeon, M. (2008). Anatomical evaluation of CT-MRI combined femoral model. *Biomed. Eng. Online*, **7**, 6.

Lemaire, T., Naili, S., and Remond, A. (2006). Multiscale analysis of the coupled effects governing the movement of interstitial fluid in cortical bone. *Biomech. Model. Mechanobiol.*, **5**(1), 39–52.

Le Novère, N., Finney, A., Hucka, M., *et al.* (2005). Minimum information requested in the annotation of biochemical models (MIRIAM). *Nat. Biotechnol.*, **23**(12), 1509–15.

Lewis, G. and Nyman, J. S. (2008). The use of nanoindentation for characterizing the properties of mineralized hard tissues: state-of-the art review. *J. Biomed. Mater. Res. B Appl. Biomater.*, **87**(1), 286–301.

Lieber, R. L. and Fridén, J. (2000). Functional and clinical significance of skeletal muscle architecture. *Muscle Nerve*, **23**(11), 1647–66.

Lord, S. R. and Menz, H. B. (2002). Physiologic, psychologic, and health predictors of 6-minute walk performance in older people. *Arch. Phys. Med. Rehabil.*, **83**(7), 907–11.

Lu, T. W. and O'Connor, J. J. (1999). Bone position estimation from skin marker co-ordinates using global optimisation with joint constraints. *J. Biomech.*, **32**(2), 129–34.

Luo, G., Ducy, P., McKee, M. D., *et al.* (1997). Spontaneous calcification of arteries and cartilage in mice lacking matrix GLA protein. *Nature*, **386**(6620), 78–81.

Manchikanti, L. Fellows, B. Pampati, V. *et al.* (2002). Comparison of psychological status of chronic pain patients and the general population. *Pain Physician*, **5**(1), 40–8.

Manfrini, M., Innocenti, M., Ceruso, M., and Mercuri, M. (2003). Original biological reconstruction of the hip in a 4-year-old girl. *Lancet*, **361**(9352), 140–2.

Marulanda, G. A., Henderson, E. R., Johnson, D. A., Letson, G. D., and Cheong, D. (2008). Orthopedic surgery options for the treatment of primary osteosarcoma. *Cancer Control*, **15**(1), 13–20.

Matsumoto, T., Ohnishi, I., Bessho, M., *et al.* (2009). Prediction of vertebral strength under loading conditions occurring in activities of daily living using a computed tomography-based nonlinear finite element method. *Spine*, **34**(14), 1464–9.

Matsuo, K. and Irie, N. (2008). Osteoclast–osteoblast communication. *Arch. Biochem. Biophys.*, **473**(2), 201–9.

Matthews, M. R. (2004). Idealisation and Galileo's pendulum discoveries: historical, philosophical and pedagogical considerations. *Sci. & Educ.*, **13**(7), 689–715.

Mayr, E. (1996). The autonomy of biology: the position of biology among the sciences. *Q. Rev. Biol.*, **71**(1), 97.

Mayr, E. (1998). *This Is Biology: The Science of the Living World.* Harvard University Press.

McNamara, L. M. (2010). Perspective on post-menopausal osteoporosis: establishing an interdisciplinary understanding of the sequence of events from the molecular level to whole bone fractures. *J. R. Soc. Interface*, **7**(44), 353–72.

Melton, L. J., 3rd, Christen, D., Riggs, B. L., *et al.* (2010). Assessing forearm fracture risk in postmenopausal women. *Osteoporos. Int.*, **21**(7), 1161–9.

Mendez, J. and Keys. A. (1960). Density and composition of mammalian muscle. *Metabolism*, **9**(2), 184–8.

Menegaldo, L. L., de Toledo Fleury, A., and Weber, H. I. (2006). A 'cheap' optimal control approach to estimate muscle forces in musculoskeletal systems. *J. Biomech.*, **39**(10), 1787–95.

Merz, B., Müller, R., and Ruegsegger, P. (1991). Solid modeling and finite element modeling of bones and implant-bone system. In *Interfaces in Medicine and Mechanics–2*, ed. K. R. Williams, A. Toni, J. Middleton, and G. Pallotti. Elsevier, pp. 345–51.

Morasso, P. (1981). Spatial control of arm movements. *Exp. Brain Res.*, **42**(2), 223–7.

Morgan, E. F., Bayraktar, H. H., and Keaveny, T. M. (2003). Trabecular bone modulus–density relationships depend on anatomic site. *J. Biomech.*, **36**(7), 897–904.

Morin, E. (1993). *Il Metodo 3. La conoscenza della conoscenza.* Raffaello Cortina Editore.

Mueller, T. L., Stauber, M., Kohler, T., *et al.* (2009) . Non-invasive bone competence analysis by high-resolution pQCT: an in vitro reproducibility study on structural and mechanical properties at the human radius. *Bone*, **44**(2), 364–71.

Muller, R. (2005). Long-term prediction of three-dimensional bone architecture in simulations of pre-, peri- and post-menopausal microstructural bone remodeling. *Osteoporos. Int.*, **16**(Suppl 2), S25–35.

Nevill, A. M., Holder, R. L., and Stewart, A. D. (2003). Modeling elite male athletes' peripheral bone mass, assessed using regional dual X-ray absorptiometry. *Bone*, **32**(1), 62–8.

Noble, B. S. (2008). The osteocyte lineage. *Arch. Biochem. Biophys.*, **473**(2), 106–11.

Noble, D. (2002). Modeling the heart – from genes to cells to the whole organ. *Science*, **295**(5560), 1678–82.

Noble, D. (2006). *The Music of Life: Biology Beyond the Genome.* Oxford University Press.

Nystrom, L. M. and Morcuende, J. A. (2010). Expanding endoprosthesis for pediatric musculoskeletal malignancy: current concepts and results. *Iowa Orthop. J.*, **30**, 141–9.

Ogata, Y. (2008). Bone sialoprotein and its transcriptional regulatory mechanism. *J. Periodontal Res.*, **43**(2), 127–35.

Ohman, C., Baleani, M., Perilli, E., *et al.* (2007). Mechanical testing of cancellous bone from the femoral head: experimental errors due to off-axis measurements. *J. Biomech.*, **40**(11), 2426–33.

Ohman, C., Dall'Ara, E., Baleani, M., Van Sint Jan, S., and Viceconti, M. (2008). The effects of embalming using a 4% formalin solution on the compressive mechanical properties of human cortical bone. *Clin. Biomech. (Bristol, Avon)*, **23**(10), 1294–8.

Öhman, C., Baleani, M., and Viceconti, M. (2009). The effect of specimen geometry on the mechanical behaviour of human cortical bone. In *IV International Conference on Computational Bioengineering (ICCB 2009), Bertinoro, Italy.*

Pakarinen, T. K., Laine, H. J., Honkonen, S. E., *et al.* (2002). Charcot arthropathy of the diabetic foot: current concepts and review of 36 cases. *Scand. J. Surg.*, **91**(2), 195–201.

Papachroni, K. K., Karatzas, D. N., Papavassiliou, K. A., Basdra, E. K., and Papavassiliou, A. G. (2009). Mechanotransduction in osteoblast regulation and bone disease. *Trends Mol. Med.*, **15**(5), 208–16.

Parfitt, A. M. (2001). The bone remodeling compartment: a circulatory function for bone lining cells. *J. Bone Miner. Res.*, **16**(9), 1583–5.

Pavol, M. J., Owings, T. M., and Grabiner, M. D. (2002). Body segment inertial parameter estimation for the general population of older adults. *J. Biomech.*, **35**(5), 707–12.

Peng, L., Bai, J., Zeng, X., and Zhou, Y. (2006). Comparison of isotropic and orthotropic material property assignments on femoral finite element models under two loading conditions. *Med. Eng. Phys.*, **28**(3), 227–33.

Phan, C. M., Macklin, E. A., Bredella, M. A., *et al.* (2010). Trabecular structure analysis using C-arm CT: comparison with MDCT and flat-panel volume CT. *Skeletal Radiol.*, **40**(8), ID 65–72.

Piscitelli, P., Brandi, M. L., Tarantino, U., *et al.* (2010). Incidenza e costi delle fratture di femore in Italia: studio di estensione 2003–2005 [Incidence and socioeconomic burden of hip fractures in Italy: extension study 2003–2005]. *Reumatismo*, **62**(2), 113–18.

Polgar, K., Gill, H. S., Viceconti, M., Murray, D. W., and O'Connor, J. J. (2003). Strain distribution within the human femur due to physiological and simplified loading: finite element analysis using the muscle standardized femur model. *Proc. Inst. Mech. Eng. H*, **217**(3), 173–89.

Polkinghorne, J. C. (2007). Reductionism. In *Interdisciplinary Encyclopedia of Religion and Science.* www.disf.org/en/Voci/104.asp. Accessed May 12th, 2011.

Popper, K. R. (1992). *The Logic of Scientific Discovery.* Routledge.

Popper, K. R. (1999). *All Life is Problem Solving.* Routledge.

Praagman, M., Chadwick, E. K., van der Helm, F. C., and Veeger, H. E. (2006). The relationship between two different mechanical cost functions and muscle oxygen consumption. *J. Biomech.*, **39**(4), 758–65.

Qin, Y. X., Lam, H., Ferreri, S., and Rubin, C. (2010). Dynamic skeletal muscle stimulation and its potential in bone adaptation. *J. Musculoskelet. Neuronal Interact.*, **10**(1), 12–24.

Queipo, N. V., Haftka, R. T., Shyy, W., *et al.* (2005). Surrogate-based analysis and optimization. *Prog. Aerosp. Sci.*, **41**(1), 1–28.

Reid, I. R. (2008). Effects of beta-blockers on fracture risk. *J. Musculoskelet. Neuronal. Interact.*, **8**(2), 105–10.

Robling, A. G. and Stout, S. D. (1999). Morphology of the drifting osteon. *Cells Tissues Organs*, **164**(4), 192–204.

Roeder, B. A., Kokini, K., Sturgis, J. E., Robinson, J. P., and Voytik-Harbin, S. L. (2002). Tensile mechanical properties of three-dimensional type I collagen extracellular matrices with varied microstructure. *J. Biomech. Eng.*, **124**(2), 214–22.

Rohrle, O. and Pullan, A. J. (2007). Three-dimensional finite element modelling of muscle forces during mastication. *J. Biomech.*, **40**(15), 3363–72.

Roschger, P., Fratzl, P., Eschberger, J., and Klaushofer, K. (1998). Validation of quantitative backscattered electron imaging for the measurement of mineral density distribution in human bone biopsies. *Bone*, **23**(4), 319–26.

Rowland, R. E. (1966). Exchangeable bone calcium. *Clin. Orthop. Relat. Res.*, **49**, 233–48.

Rubin, L. A., Hawker, G. A., Peltekova, V. D., *et al.* (1999). Determinants of peak bone mass: clinical and genetic analyses in a young female Canadian cohort. *J. Bone Miner. Res.*, **14**(4), 633–43.

Rubin, M. A. and Jasiuk, I. (2005). The TEM characterization of the lamellar structure of osteoporotic human trabecular bone. *Micron*, **36**(7–8), 653–64.

Sackett, D. L., Rosenberg, W. M., Gray, J. A., Haynes, R. B., and Richardson, W. S. (1996). Evidence based medicine: what it is and what it isn't. *BMJ*, **312**(7023), 71–2.

Sadowsky, O., Chintalapani, G., and Taylor R. H. (2007). Deformable 2D-3D registration of the pelvis with a limited field of view, using shape statistics. *Med. Image Comput. Comput. Assist. Interv.*, **10**(2), 519–26.

Sai, A. J., Gallagher, J. C., Smith, L. M., and Logsdon, S. (2010). Fall predictors in the community dwelling elderly: a cross sectional and prospective cohort study. *J. Musculoskelet. Neuronal Interact.*, **10**(2), 142–50.

Schiessl, H., Frost, H. M., and Jee, W. S. (1998). Estrogen and bone-muscle strength and mass relationships. *Bone*, **22**(1), 1–6.

Schileo, E., Taddei, F., Malandrino, A., Cristofolini, L., and Viceconti, M. (2007). Subject-specific finite element models can accurately predict strain levels in long bones. *J. Biomech.*, **40**(13), 2982–9.

Schmid, J. and Magnenat-Thalmann, N. (2008). MRI bone segmentation using deformable models and shape priors. *Med. Image. Comput. Comput. Assist. Interv.*, **11**(1), 119–26.

Scholz J. P. and Schoner, G. (1999). The uncontrolled manifold concept: identifying control variables for a functional task. *Exp. Brain Res.*, **126**(3), 289–306.

Schulte, F. A., Lambers, F. M., Kuhn, G., and Muller, R. (2010). In vivo micro-computed tomography allows direct three-dimensional quantification of both bone formation and bone resorption parameters using time-lapsed imaging. *Bone*, **48**(3), 433–42.

Seireg, A. and Arvikar, R. J. (1975). The prediction of muscular load sharing and joint forces in the lower extremities during walking. *J. Biomech.*, **8**(2), 89–102.

Serway, R. A. (1995) . Coefficients of friction. *Physics for Scientists and Engineers* 4th edn., p. 126. www.physlink.com/reference/FrictionCoefficients.cfm. Accessed May 13th, 2011.

Settle, T. (1989). Van Rooijen and Mayr versus Popper: is the universe causally closed? *Br. J. Philos. Sci.*, **40**(3), 398–403.

Shea, J. E. and Miller, S. C. (2005). Skeletal function and structure: implications for tissue-targeted therapeutics. *Adv. Drug Deliv. Rev.*, **57**(7), 945–57.

Shen, Z. L., Dodge, M. R., Kahn, H., Ballarini, R., and Eppell, S. J. (2008). Stress–strain experiments on individual collagen fibrils. *Biophys. J.*, **95**(8), 3956–63.

Shirazi-Adl, A. (1994). Analysis of role of bone compliance on mechanics of a lumbar motion segment. *J. Biomech. Eng.*, **116**(4), 408–12.

Sievanen, H. (2010). Immobilization and bone structure in humans. *Arch. Biochem. Biophys.*, **503**(1), 146–52.

Silver, F. H. (2009). The importance of collagen fibers in vertebrate biology. *J. Eng.Fiber. Fabr.*, **4**(2), 9–17.

Singh, P. K., Marzo, A., Howard, B., *et al.* (2010). Effects of smoking and hypertension on wall shear stress and oscillatory shear index at the site of intracranial aneurysm formation. *Clin. Neurol. Neurosurg.*, **112**(4), 306–13.

Skedros, J. G., Mendenhall, S. D., Kiser, C. J., and Winet, H. (2009). Interpreting cortical bone adaptation and load history by quantifying osteon morphotypes in circularly polarized light images. *Bone*, **44**(3), 392–403.

Skerry, T. M. (1997). Mechanical loading and bone: what sort of exercise is beneficial to the skeleton? *Bone*, **20**(3), 179–81.

Skerry T. M. (2008). The response of bone to mechanical loading and disuse: fundamental principles and influences on osteoblast/osteocyte homeostasis. *Arch. Biochem. Biophys.*, **473**(2), 117–23.

Standal, T., Borset, M., and Sundan, A. (2004). Role of osteopontin in adhesion, migration, cell survival and bone remodeling. *Exp. Oncol.*, **26**(3), 179–84.

STEP Consortium (2007). *Seeding the EuroPhysiome: A Roadmap to the Virtual Physiological Human*. www.europhysiome.org/roadmap. Accessed May 12th, 2011.

Strang, G. and Fix, G. (2008). *An Analysis of the Finite Element Method*. 2nd edn. Wellesley.

Strasser, S., Zink, A., Janko, M., Heckl, W. M., and Thalhammer, S. (2007). Structural investigations on native collagen type I fibrils using AFM. *Biochem. Biophys. Res. Commun.*, **354**(1), 27–32.

Tabor, Z. and Rokita, E. (2007). Quantifying anisotropy of trabecular bone from gray-level images. *Bone*, **40**(4), 966–72.

Taddei, F., Viceconti, M., Manfrini, M., and Toni, A. (2003). Mechanical strength of a femoral reconstruction in paediatric oncology: a finite element study. *Proc. Inst. Mech. Eng. H*, **217**(2), 111–19.

Taddei, F., Stagni, A., Cappello, M., *et al.* (2005). Kinematic study of a reconstructed hip in paediatric oncology. *Med. Biol. Eng. Comput.*, **43**(1), 102–6.

Taddei, F., Martelli, S., Reggiani, B., Cristofolini, L., and Viceconti, M. (2006). Finite-element modeling of bones from CT data: sensitivity to geometry and material uncertainties. *IEEE Trans. Biomed. Eng.*, **53**(11), 2194–200.

Taddei, F., Schileo, E., Helgason, B., Cristofolini, L., and Viceconti, M. (2007). The material mapping strategy influences the accuracy of CT-based finite element models of bones: an evaluation against experimental measurements. *Med. Eng. Phys.*, **29**(9), 973–9.

Taddei, F., Pani, M., Zovatto, L., Tonti, E., and Viceconti, M. (2008). A new meshless approach for subject-specific strain prediction in long bones: evaluation of accuracy. *Clin. Biomech. (Bristol, Avon)*, **23**(9), 1192–9.

Taghavi, R. (1996). Automatic, parallel and fault tolerant mesh generation from CAD. *Eng. Comput.*, **12**(3), 178–85.

Takeda, S., Elefteriou, F., Levasseur, R., *et al.* (2002). Leptin regulates bone formation via the sympathetic nervous system. *Cell*, **111**(3), 305–17.

Takeuchi, T. and Gonda, T. (2004). Distribution of the pores of epithelial basement membrane in the rat small intestine. *J. Vet. Med. Sci.*, **66**(6), 695–700.

Tang, Y., Ballarini, R., Buehler, M. J., and Eppell, S. J. (2010). Deformation micromechanisms of collagen fibrils under uniaxial tension. *J. R. Soc. Interface*, **7**(46), 839–50.

Tassani, S., Ohman, C., Baleani, M., Baruffaldi, F., and Viceconti, M. (2010). Anisotropy and inhomogeneity of the trabecular structure can describe the mechanical strength of osteoarthritic cancellous bone. *J. Biomech.* **43**(6), 1160–6.

Tassani, S., Ohman, C., Baruffaldi, F., Baleani, M., and Viceconti, M. (2011). Volume to density relation in adult human bone tissue. *J. Biomech.*, **44**(1), 103–8.

Taylor C. F., Field, D., Sansone, S. A., *et al.* (2008). Promoting coherent minimum reporting guidelines for biological and biomedical investigations: the MIBBI project. *Nat. Biotechnol.*, **26**(8), 889–96.

Testi, D., Zannoni, C., Cappello, A., and Viceconti, M. (2001). Border-tracing algorithm implementation for the femoral geometry reconstruction. *Comput. Methods Programs Biomed.*, **65**(3), 175–82.

Testi, D., Quadrani, P., and Viceconti, M. (2010). PhysiomeSpace: digital library service for biomedical data. *Philos. Transact. A Math. Phys. Eng. Sci.*, **368**(1921), 2853–61.

Thelen, D. G., Anderson, F. C., and Delp S. L. (2003). Generating dynamic simulations of movement using computed muscle control. *J. Biomech.*, **36**(3), 321–8.

Theoleyre, S., Wittrant, Y., Tat, S. K., *et al.* (2004). The molecular triad OPG/RANK/RANKL: involvement in the orchestration of pathophysiological bone remodeling. *Cytokine Growth Factor Rev.*, **15**(6), 457–75.

Thiel, R., Stroetmann, K. A., Stroetmann, V. N., and Viceconti, M. (2009). Designing a socio-economic assessment method for integrative biomedical research: the Osteoporotic Virtual Physiological Human project. *Stud. Health Technol. Inform.*, **150**, 876–80.

Todorov, E. and Jordan, M. I. (2002). Optimal feedback control as a theory of motor coordination. *Nat. Neurosci.*, **5**(11), 1226–35.

Tonti, E. (2001). A direct discrete formulation of field laws: the cell method. *Comput. Model. Eng. Sci.*, **2**(2), 237–58.

Truc, P. T., Kim, T. S., Lee, S., and Lee Y. K. (2009). A study on the feasibility of active contours on automatic CT bone segmentation. *J. Digit. Imaging.*, **22**(6), 793–805.

Tsirakos, D., Baltzopoulos, V., and Bartlett, R. (1997). Inverse optimization: functional and physiological considerations related to the force-sharing problem. *Crit. Rev. Biomed. Eng.*, **25**(4–5), 371–407.

Turner, C. H. and Robling, A. G. (2005). Mechanisms by which exercise improves bone strength. *J. Bone Miner. Metab.*, **23**(Suppl), 16–22.

van der Linden, J. C., Day, J. S., Verhaar, J. A., and Weinans, H. (2004). Altered tissue properties induce changes in cancellous bone architecture in aging and diseases. *J. Biomech.*, **37**(3), 367–74.

van der Plas, A. and Nijweide, P. J. (1992). Isolation and purification of osteocytes *J. Bone. Miner. Res.*, **7**(4), 389–96.

van Lenthe, G. H., Mueller, T. L., Wirth, A. J., and Muller, R. (2008). Quantification of bone structural parameters and mechanical competence at the distal radius. *J. Orthop. Trauma*, **22**(8 Suppl), S66–72.

van Oers, R. F., Ruimerman, R., Tanck, E., Hilbers, P. A., and Huiskes, R. (2008). A unified theory for osteonal and hemi-osteonal remodeling. *Bone*, **42**(2), 250–9.

Van Rietbergen, B., Weinans, H., Huiskes, R., and Polman, B. J. W. (1996). Computational strategies for iterative solutions of large FEM applications employing voxel data. *Int. J. Numer. Methods Eng.*, **39**(16), 2743–67.

Van Sint Jan, S. (2007). *Color Atlas of Skeletal Landmark Definitions: Guidelines for Reproducible Manual and Virtual Palpations.* Churchill Livingstone.

Van Sint Jan, S., Sobzack, S., Dugailly, P. M., *et al.* (2006). Low-dose computed tomography: a solution for in vivo medical imaging and accurate patient-specific 3D bone modeling? *Clin. Biomech. (Bristol, Avon)*, **21**(9), 992–8.

Viceconti, M. and Clapworthy, G. (2006). The virtual physiological human: challenges and opportunities. In *3rd IEEE International Symposium on Biomedical Imaging: Macro to Nano*, pp. 812–15.

Viceconti, M. and Kohl, P. (2010). The virtual physiological human: computer simulation for integrative biomedicine I. *Philos. Transact. A Math. Phys. Eng. Sci.*, **368**(1920), 2591–4.

Viceconti, M. and Taddei F. (2003). Automatic generation of finite element meshes from computed tomography data. *Crit. Rev. Biomed. Eng.*, **31**(1–2), 27–72.

Viceconti, M., Bellingeri, L., Cristofolini, L., and Toni, A. (1998). A comparative study on different methods of automatic mesh generation of human femurs. *Med. Eng. Phys.*, **20**(1), 1–10.

Viceconti, M., Olsen, S., Nolte, L. P., and Burton, K. (2005). Extracting clinically relevant data from finite element simulations. *Clin. Biomech. (Bristol, Avon)*, **20**(5), 451–4.

Viceconti, M., Brusi, G., Pancanti, A., and Cristofolini, L. (2006a). Primary stability of an anatomical cementless hip stem: a statistical analysis. *J. Biomech.*, **39**(7), 1169–79.

Viceconti, M. Testi, D. Taddei, F. *et al.* (2006b). Biomechanics Modeling of the Musculoskeletal Apparatus: Status and Key Issues. *Proceedings of the IEEE*, **94**(4), 725–39.

Viceconti, M., Taddei, F., Montanari, L., *et al.* (2007). Multimod Data Manager: a tool for data fusion. *Comput. Methods Programs Biomed.*, **87**(2), 148–59.

Viceconti, M., Clapworthy, G., and Van Sint Jan, S. (2008a). The Virtual Physiological Human – a European initiative for in silico human modelling. *J. Physiol. Sci.*, **58**(7), 441–6.

Viceconti, M., Taddei, F., Van Sint Jan, S., *et al.* (2008b). Multiscale modelling of the skeleton for the prediction of the risk of fracture. *Clin. Biomech. (Bristol, Avon)*, **23**(7), 845–52.

Viceconti, M., Clapworthy, G., Testi, D., Taddei, F., and McFarlane, N. (2010). Multimodal fusion of biomedical data at different temporal and dimensional scales. *Comput. Methods Programs Biomed.*, **102**(3), 227–37.

Wang, J. H. and Thampatty, B. P. (2006). An introductory review of cell mechanobiology. *Biomech. Model. Mechanobiol.*, **5**(1), 1–16.

Wang Y. C. and Chen B. S. (2010). Integrated cellular network of transcription regulations and protein-protein interactions. *BMC Syst Biol.*, **4**, 20.

Watts, N. B. (1999). Clinical utility of biochemical markers of bone remodeling. *Clin. Chem.*, **45**(8 Pt 2), 1359–68.

Webster, D., Wasserman, E., Ehrbar, M., *et al.* (2010). Mechanical loading of mouse caudal vertebrae increases trabecular and cortical bone mass-dependence on dose and genotype. *Biomech. Model. Mechanobiol.*, **9**(6), 737–47.

Weiner, S. and Traub, W. (1992). Bone structure: from angstroms to microns. *FASEB J.*, **6**(3), 879–85.

Weiner, S., Arad, T., Sabanay, I., and Traub, W. (1997). Rotated plywood structure of primary lamellar bone in the rat: orientations of the collagen fibril arrays. *Bone*, **20**(6), 509–14.

Weiss, L. (1988). *Cell and Tissue Biology: A Textbook of Histology*. Urban & Schwarzenberg.

Wikipedia. (2008). Reductionism. *Wikipedia: The Free Encyclopedia*. http://en.wikipedia. org/wiki/Reductionism. Accessed May 12th, 2011.

Winby, C. R., Lloyd, D. G., and Kirk, T. B. (2008). Evaluation of different analytical methods for subject-specific scaling of musculotendon parameters. *J. Biomech.*, **41**(8), 1682–8.

Wu, C. F. J. (1986). Jackknife, bootstrap and other resampling methods in regression analysis. *Ann. Stat.*, **14**(4), 1261–95.

Wu, G., Siegler, S., Allard, P., *et al.* (2002). ISB recommendation on definitions of joint coordinate system of various joints for the reporting of human joint motion – part I: ankle, hip, and spine. *J. Biomech.*, **35**(4), 543–8.

Wu, G., van der Helm, F. C., Veeger, H. E., *et al.* (2005). ISB recommendation on definitions of joint coordinate systems of various joints for the reporting of human joint motion – Part II: shoulder, elbow, wrist and hand. *J. Biomech.*, **38**(5), 981–92.

Yamaguchi, F. (2002). *Computer-Aided Geometric Design: A Totally Four-Dimensional Approach*. Springer Verlag.

Yamaguchi, G. A. (1990). A survey of human musculotendon actuator parameters. In *Multiple Muscle Systems: Biomechanics and Movement Organization,* ed. J. M. Winters and S. L.-Y. Woo. Springer-Verlag.

Yamaguchi, G. T. (2001). *Dynamic Modeling of Musculoskeletal Motion: A Vectorized Approach for Biomechanical Analysis in Three Dimensions*. Springer Verlag.

Yamashita, J., Li, X., Furman, B. R., *et al.* (2002). Collagen and bone viscoelasticity: a dynamic mechanical analysis. *J. Biomed. Mater. Res.*, **63**(1), 31–6.

Yang, L., van der Werf, K. O., Fitie, C. F., *et al.* (2008). Mechanical properties of native and cross-linked type I collagen fibrils. *Biophys. J.*, **94**(6), 2204–11.

Yao, J., Salo, A. D., Lee, J., and Lerner, A. L. (2008). Sensitivity of tibio-menisco-femoral joint contact behavior to variations in knee kinematics. *J. Biomech.*, **41**(2), 390–8.

Yoon, Y. J. and Cowin, S. C. (2008). The estimated elastic constants for a single bone osteonal lamella. *Biomech. Model. Mechanobiol.*, **7**(1), 1–11.

Yosibash, Z., Tal, D., and Trabelsi, N. (2010). Predicting the yield of the proximal femur using high-order finite-element analysis with inhomogeneous orthotropic material properties. *Philos. Transact. A Math. Phys. Eng. Sci.*, **368**(1920), 2707–23.

Yu, H., Wooley, P. H., and Yang, S.-Y. (2009). Biocompatibility of poly-epsilon-caprolactone-hydroxyapatite composite on mouse bone marrow-derived osteoblasts and endothelial cells. *J. Orthop. Surg. Res.*, **4**, 5.

Yucesoy, C. A., Koopman, B. H., Baan, G. C., Grootenboer H. J., and Huijing, P. A. (2003). Effects of inter- and extramuscular myofascial force transmission on adjacent synergistic muscles: assessment by experiments and finite-element modeling. *J. Biomech.*, **36**(12), 1797–811.

Zajac, F. E. (1989). Muscle and tendon: properties, models, scaling, and application to biomechanics and motor control. *Crit. Rev. Biomed. Eng.*, **17**(4), 359–411.

Zannoni, C., Cappello, A. and Viceconti, M. (1998). Optimal CT scanning plan for long-bone 3-D reconstruction. *IEEE Trans. Med. Imaging*, **17**(4), 663–6.

Zienkiewicz, O. C. and Taylor, R. L. (2005). *The Finite Element Method for Solid and Structural Mechanics*. 6th edn. Butterworth-Heinemann.

Zovatto, L. and Nicolini, M. (2003). The meshless approach for the cell method: a new way for the numerical solution of the discrete conservation laws. *Int. J. Comput. Methods Eng. Sci. Mech.*, **4**(4), 869–80.

Zovatto, L. and Nicolini, M. (2006). Extension of the meshless approach for the cell method to three-dimensional numerical integration of discrete conservation laws. *Int. J. Comput. Methods Eng. Sci. Mech.*, **7**(2), 69–79.

Index